PRO BIOTICS

How, in our polluted world, friendly bacteria can help your body to perform at an optimal level.

The front cover illustration shows *Lactobacillus bulgaricus* from live yogurt under the microscope (Science Photo Library)

PRO BIOTICS

How to use 'friendly bacteria' to restore total health and vitality

LEON CHAITOW N.D., D.O.
and NATASHA TRENEV

Thorsons
An Imprint of HarperCollins*Publishers*

Thorsons
An Imprint of HarperCollins*Publishers*
77–85 Fulham Palace Road,
Hammersmith, London W6 8JB

Published by Thorsons 1990
7 9 10 8 6

A catalogue record for this book is available from the British Library

ISBN 0 7225 1919 2

Printed in Great Britain by
Mackays of Chatham PLC, Chatham, Kent

Contents

Acknowledgements

I could not have written this book without access to the vast amount of knowledge on the subject of the friendly bacteria of my co-author, Natasha Trenev. She and her dedicated staff made research easy and even a pleasure, I wish to thank her, her husband Yordan, and all those who made my time in Los Angeles so pleasant and instructive.

I wish also to thank Dr Jeremija Rašić for his interest and supportive advice in areas of study in which no-one on this planet knows more. Finally to our publishers special thanks for their confidence in this subject, one which has been neglected but which needs to be brought to the very centre of attention of all who value health.

I dedicate this book to Alkmini with love.

Leon Chaitow

I dedicate this book to my husband, Yordan Trenev, who has always inspired me to do my best.

Natasha Trenev

WHAT IS A BILLION?

The word billion in Europe means something other than the same word in the USA. This is because an American billion is equal to a thousand million, whereas the European billion equals a million million.

The number of times the number 10 is multiplied by itself is another way of expressing such astronomical figures, 10×10 equals 100. This can also be written 10^2. That makes a thousand 10^3 and 10,000 10^4. Expressed mathematically, therefore, an American billion is 10^9, whereas in Europe it is 10^{12}.

Similarly a trillion in America is 10^{12} and in Europe it is 10^{18}.

In our text we are using the American method of counting and so any time we refer to a billion, we mean a thousand million.

Publisher's Note

The book *Probiotics* contains facts and information concerning body ecology. The opinions expressed are those of the authors, and may differ from those of other experts. Nothing in the publication shall be construed as a representation concerning any product whatsoever.

Foreword

I am pleased to introduce this very informative book about probiotics to all who are interested in health and nutrition.

The word 'probiotic' derives from two Greek words meaning 'for life' and refers to organisms such as friendly bacteria which contribute so much to the health and balance of the intestinal tract, thus benefiting us by protecting against disease and improving nutrition.

The regular inclusion of friendly bacteria (lactobacilli, bifidobacteria etc.) in the diet produces beneficial effects, but since different species differ considerably in the benefits they offer, only selected strains with desirable properties should be selected for supplementation.

Those which are best are freeze-dried to preserve the viability of the organisms. To be really effective freeze dried supplements should fulfil some basic requirements such as:

(a) incorporating only the declared species and desirable strains

(b) having large numbers of viable friendly bacteria in concentrations of 10^8 to 10^9 per gram,

(c) be kept refrigerated.

The many beneficial potentials of the friendly bacteria are outlined in this easily read and understood book, with many illustrative examples. Early chapters describe the role and nature of the friendly bacteria in human nutrition, as well as their preventive and therapeutic roles in many forms of human illness.

Special attention is given to describing their antibiotic, antiviral and antifungal powers in a variety of infections, including the intestinal, vaginal, urinary tracts and other regions.

There are chapters dealing with their cholesterol fighting and their anti-carcinogenic and anti-tumour potentials, as well as their role in the management of liver disease and stress conditions. Their influence on infant nutrition is also covered.

Emphasis is put on the importance of the use of supplemental friendly bacteria using large numbers of viable organisms from selected strains.

I hope this book proves of great value to both the general reader, nutritionists and medical workers engaged in the promotion of health.

Jeremija Lj. Rašić Ph.D.
Novi Sad Institute
Yugoslavia

Chapter 1

The role and nature of friendly bacteria

In the coming pages we will together explore a dark, damp world in which a vast number of bacterial creatures live in healthy, helpful, co-operative symbiosis with their human hosts; that means all of us, at least when we are ourselves in good health.

In examining the remarkable, indeed vital, roles these friendly bacteria play in our lives we will also come across a less desirable batch of characters — the pathogenic, or disease producing, bacteria (as well as viruses, fungi etc.) which often invade certain recesses of our bodies. Our level of health depends to a large extent upon the condition of the friendly bacteria and the control which they are able to exert over their less friendly brethren, who can become opportunistically and destructively active (for us that is) when conditions are desirable for them (for example, when we have a lowered immune function).

It is true to say that our very lives depend upon the friendly bacteria with which mankind has co-existed for millions of years, and that the level of well-being we enjoy is very much involved with their having a good, wholesome environment in which to live.

The secret world within you

It has been estimated that over 400 species of bacteria inhabit our digestive tracts. Different species are found in various regions — stomach, jejunum etc. which have varying degrees of acidity relating to their function when we are in good health.

It often comes as a surprise for us to learn that the thousands of billions of bacteria living in the human gastro-intestinal tract weigh an amazing 3½ pounds (approximately 1½ kilos)! Indeed, roughly one third of the dry weight of faecal material (with water removed) consists of dead and viable bacteria. Amazingly, therefore, this population of bacteria (friendly and otherwise) far exceeds the total number of tissue cells which make up your body.

Not all bacterial inhabitants of the body are friendly, and we should distinguish between those with which we have established a symbiotic, mututally beneficial, relationship over the aeons of evolutionary time and those which are hostile with the potential for causing distress and disease.

A complete bacteriological analysis of one tiny sample of human faeces could take a year or more to accomplish, which is an indication of the complexity of this vast ecosystem that lies within us all. Some of the friendly bacteria can only thrive in an environment almost totally free of oxygen (such as bifidobacteria, an extremely useful inhabitant, as we shall see later) whereas others require small amounts of oxygen just as we do (and are therefore called microaerophilic organisms) in order to live and thrive (e.g. *Lactobacillus acidophilus*, which also has some strains which can survive without oxygen, and versatile strains which are both aerobic and anaerobic.)

In the study of these organisms it has been found useful to distinguish between those which are a normal part of almost everyone's bowel and those which are more or less transient nomads, colonizing an area for a while and then disappearing.

What disturbs the balance?

Some of these transients are extremely useful, as will become clear. Other, less welcome, interlopers usually only find a niche to occupy when there is some disturbance of the local ecology. Such a disturbance may occur when medical antibiotics, e.g. penicillin, are used in treating an infection, since one side-effect of penicillin is that it also kills large numbers of the normal (friendly) bacterial inhabitants of the body.

Another disturbing influence on the bowel ecology would be low levels of hydrochloric acid in the digestive juices, a condition which is very common as we grow older, and not uncommon in young people, especially those prone to allergic conditions such as asthma.

Altered degrees of acidity anywhere in the bowel (different regions require differing levels of acidity/alkalinity) will change the nature of the habitat for bacteria, thus altering the type, quantity and behaviour (production of enzymes etc.) of micro-organisms found there.

The renowned medical researcher René Dubos has found that when animals are psychologically stressed the health of their intestinal flora is adversely affected. He maintains that the factors which most influence the intestinal flora (the term used to describe the host of bacterial cultures living there) are the diet of the individual, the use of antibiotic drugs and stress of various sorts.

Studies show that the lactobacilli are the friendly bacteria which are most affected by stress.

When do these 'squatters' arrive?

Within a few days of birth the gastro-intestinal tract is colonized by bacteria, and the pattern of birth strongly influences the pattern of colonization. For example *Bifidobacterium infantis* was found in around 60 per cent of 4–6 day old babies born vaginally, at full-term, all of whom were found to be colonized by bacteria which do not require oxygen (anaerobic). In contrast only 9 per cent of babies born by Caesarean section were so colonized and a mere 9 per cent of these comprised *infantis*.

Breast feeding also alters the pattern of the type of bacterial colonization which occurs in babies, with fewer non-bifidobacteria found in breast-fed babies when compared with those who have been formula-fed.

It has been possible to determine the direction in which this colonization takes place by examining the bowels of babies with congenital obstructive conditions. This shows that in the regions above the obstruction colonization had taken place, whereas the region beyond the blockage remained bacteria free.

It is clear, therefore, that the invaders enter the system from the mouth, and so whatever food is eaten will strongly influence the nature of the initial colonization process.

What benefits do we derive from these useful but uninvited 'house-guests'?

In return for the living space and nutrients which we provide for the friendly bacteria we have exacted a price, and they certainly pay their way. In the 'trade-off':

- They manufacture B-vitamins, such as biotin, niacin (B3), pyridoxine (B6) and folic acid.
- By providing the enzyme lactase they enhance, and indeed allow, the digestion of milk-based foods, and the vital calcium which they contain, for people who cannot otherwise digest milk.
- They predigest the protein of the cultured milk/yogurt in which they are often found, thus enhancing protein digestion and absorption.
- They act as anti-carcinogenic (anti-cancer) factors, with powerful anti-tumour potentials.
- They act as 'watchdogs' by keeping an eye on, and effectively controlling, the spread of undesirable micro-organisms (by altering the acidity of the region they inhabit and/or producing specific antibiotic substances, as well as by depriving rival unfriendly bacteria of their nutrients). The antibiotics some of the friendly bacteria produce are effective against many harmful bacteria, viruses and fungi. Not the least of the potentially harmful yeasts controlled by some lactobacilli is *Candida albicans*, now implicated in many health problems in people who are malnourished or whose immune systems are depleted. Food poisoning and many bowel and urinary tract infections (diarrhoea, cystitis etc.) can be prevented and treated using high potency bacterial cultures.
- They help considerably to enhance bowel function. Where bowel bacteria are absent, the function of peristalsis is impaired, and the amount of time it takes for food to pass completely through the system is much increased.
- They effectively help to control high cholesterol levels, thereby affording us protection from the cardiovascular damage which excessive levels of this nevertheless important substance can create.
- They sometimes act to relieve the symptoms of anxiety.
- They have been shown to control facial acne in 80 per cent of adolescents with this problem.
- They play a vital role in the development of a healthy digestive tract in babies. Bifidobacteria in particular are vital for this job and are frequently supplemented in order to enhance this function in non-breast-fed babies who may

have bowel and absorption problems. As will become clear in Chapter 13, modern breast-fed babies are now less well endowed with *B. infantis* than in the recent past, due, it is thought, to environmental pollution affecting them and their mothers. Bifidobacteria supplementation, therefore, has been shown to benefit breast-fed babies as well.

- They play a role in protecting against the negative effects of radiation and toxic pollutants, enhancing immune function.
- They have been shown in various studies to be useful in the treatment of such diverse conditions as psoriasis, eczema, allergies, migraine, gout, rheumatic and arthritic conditions, as well as those mentioned above such as cancer, skin complaints, cystitis and many bowel problems, including colitis and irritable bowel syndrome.

Why we use the word 'Probiotics'

Friendly bacteria, therefore, have the potential to play twin roles in their relationship with us, and we should make a distinction between these. First, they can markedly enhance our nutritional status and, second, they have specific and important therapeutic roles.

It is for these multiple and interacting benefits which friendly bacteria bestow on us that the term 'probiotics' has been coined. This indicates that they support and enhance life: our life, and their own, in contrast to the anti-life activity of antibiotics which indiscriminately kill friend and foe alike when they are administered.

In the coming chapters each of these elements will be explored in detail allowing us to realize the many ways in which we can employ the supplementation of various different cultures to great advantage.

Not all 'friendly' bacteria are the same

However, an important caution is that not all the friendly bacteria play all the various roles mentioned, nor indeed do all strains of the same culture have the same potentials. For example, it has been found that some strains of *Lactobacillus acidophilus* produce the antibiotic acidophilin (strain DDS 1 for instance) when they are grown in a milk medium, while many other *Lactobacillus acidophilus* cultures produce no acidophilin at all. Just as there are different qualities in varying strains of potatoes or dogs, so there may be very different characteristics in bacterial cultures of the same family, genus and species. This distinction will become important when we consider the possible therapeutic use of particular strains.

Variations when health and nutrition status alters

We certainly need to distinguish between the work these minute creatures do when their colonies are healthily and productively engaged in useful tasks, when we are in good health with a sound bowel ecology, and the way they act when they are introduced to our bodies in the form of food (yogurt, cultured milk such as kefir etc.) or in the form of high potency cultures (powders, capsules etc.) when we are not in such good health. Since these bacteria live inside us it is not surprising that their own health is directly influenced by the diet we consume, both in type and quality.

For example, bifidobacteria increase in number on a high complex-carbohydrate lacto-vegetarian diet, whilst another group, the bacteroides bacteria, flourish when the diet is rich in meat and fat.

It has been shown that the diet of the host has specific effects on the type of enzymes the bacteria in the bowel produce, as well as on the numbers of the various types of bacteria. All these changes influence the potential for good which is possible when the bacteria are well nourished and unstressed.

Commercial influences

We also need to be critically aware of the way commercial production of many cultured foods (yogurt etc.) and supplemental forms of various lactobacilli can damage their effectiveness, since it is becoming increasingly clear that many such 'foods' are almost totally devoid of active cultures. Some products contain only minimal amounts of the cultures the consumer may believe to be present in them, or have actually never even been in contact with them.

This is true of some yogurts and some supplemental forms of friendly bacteria. It is as important to identify the potentials for health enhancement which good products provide as it is to identify the phony, and in some instances the outright dangerous.

As Beverly Friend and Khem Shahani Ph.D., two of the most renowned researchers in this area of science, so clearly put it: 'Commercial preparations of cultures (such as centrifugation, see Chapter 16) shown to be effective in the laboratory may not contain sufficient viable organisms to be of benefit. Specialized media and techniques are required to handle, propagate and concentrate the lactobacilli to ensure that adequate numbers of viable organisms survive the concentration, freeze-drying or drying process needed to prepare a stable commercial product.'

What makes a friendly bacteria useful

In order for a particular organism to be considered as desirable for use in either food or supplementation, several characteristics need to be established:

1. The organism should be a normal inhabitant of, or regular and useful transient 'visitor' to the digestive tract, such as one of the lactobacilli (*Lactobacillus acidophilus*, a resident, or *L. bulgaricus*, a transient visitor).
2. After being swallowed they should be capable of surviving the digestive processes of the upper digestive tract. Therefore, proper directions to the consumer as to when to take such supplements are essential.
3. They should also be capable of surviving and growing in the intestinal regions of the tract. This calls for their being able to survive in the presence of bile acids and being able to colonize and attach to the surfaces of the gut they inhabit.
4. They should be shown to produce beneficial effects when present in the digestive tract.
5. They should remain viable and active in the food (or capsule, container etc.) in which they are present before consumption.

These criteria are all met when GOOD quality products, containing healthy, vital and viable *Lactobacillus acidophilus, Bifidobacterium bifidum* and *Lactobacillus bulgaricus* (a transient bacteria of considerable value) are consumed. There are many other desirable cultures and these will be discussed as we progress. However, there are also some which do not meet all the requirements mentioned above, and we will discuss these as well in the appropriate sections of the book.

It is now time to meet the main players in the cast of characters we have come to examine; the friendly bacteria.

Chapter 2

Meet the friendly bacteria

We will conduct you on a brief tour through the gastro-intestinal (GI) tract in order to meet the major friendly microbial flora present in each region. This will allow us also to come to understand some of the important characteristics of the leading inhabitants of this area in greater detail. The ones which interest us most, because of their well documented beneficial functions in health protection and therapy, are *Lactobacillus acidophilus, Bifidobacteria, Lactobacillus bulgaricus* and *Streptococcus thermophilus* (the last two being important yogurt cultures). We will also introduce you to a number of minor role players in the great drama of life which incessantly unfolds inside each of us. In case you are not already familiar with this region we suggest that you refer to page 20 which illustrates the various regions of the GI tract.

All of life is here

Within these regions battles rage; populations rise and fall, affected just as we are by local environmental conditions; industry thrives and constant defence is exercised against interlopers and dangerous aliens who may enter unannounced; colonists roam and settle — some permanently, some only briefly; in general we have in miniature many of terrestrial life's vicissitudes, problems and solutions.

The most useful of the inhabitants of the GI tract take great care (albeit unknowingly) of their environment (the human body) which so generously provides them with shelter, warmth and food, in return for essential labour and defence roles.

Not only do these guests help to digest our food, manufacture essential vitamins and prevent cancer-causing chemicals from forming, but they also act strongly to defend against undesirable micro-organisms which may appear. For there are opportunistic vagabonds, such as *Candida albicans*, forever lurking in the hope of spreading from the small but permanent niche in which they survive inside each of us. They will spread aggressively when the control exercised by friendly bacteria is weakened due to ecological disaster, such as the administration of antibiotics.

The major inhabitants of the GI tract

A. Starting at the beginning of the gastro-intestinal tract, *the mouth*, we find that the saliva contains a motley assortment of microbes, with streptococci, lactobacilli, veillonella, bacteroides, fusobacteria, staphylococci, and cornybacteria usually all present in varying degrees. Each millilitre of saliva contains between 10,000 and one billion (10^9) micro-organisms.

B. Moving on to *the stomach* we find very few bacteria because of the presence, during digestion, of a host of enzymes and acids. When no food is present in the stomach, the levels of bacteria found are very low indeed, either none at all or no more than 1,000 per millilitre of contents. Those micro-organisms which are swallowed with food, or in saliva, are rapidly killed by hydrochloric acid, or if they are acid-resistant and survive, they are swiftly moved on to the small intestine by the peristaltic waves which keep the contents of the GI tract moving. The acidity of a normal stomach during digestion is around pH3 which kills most bacteria. Some evidence exists for pH going below 1.0 to facilitate optimal digestion. (See panel below for explanation of acidity measurement.)

C. *The small intestine* is the region between the stomach and the large intestine and is a major area for food digestion. It is itself divided into the duodenum, which receives juices from the pancreas and the liver (gall bladder), the jejunum and the ileum. The small intestine is a 'zone of transition' between the sparsely populated stomach and the luxuriant flora of the large intestine. In the upper segments of the small intestine only small numbers (around 10,000 organisms per millilitre) of usually transient bacterial colonization are found, but lower down, in the ileum for example, we find a rich and permanent population.

The types found here are lactobacilli, streptococci, and small

ACIDITY

Acidity is an important environmental control factor influencing bacterial colonization and destruction of undesirable interlopers.

The letters pH (which stands for Hydrogen potential) indicate the level of acidity, and there is a standard scale in which anything below the number of pH7 indicates acid, above pH7 is alkaline, and pH7 itself is neutral.

Some bifidobacteria require a moderately acid environment, and the stools of new-born, breast-fed babies with a high content of bifidobacteria actually range between pH5 and 5.5. Different species (as listed in this chapter) require a pH of anything from 5.5 to 7 and growth and activity may cease when these parameters are exceeded.

There are also temperature levels which are ideal for their growth and function which if exceeded in either direction (too hot or too cold) will cause them to cease functioning.

The levels of local acidity change in bottle-fed infants, older individuals and during weaning.

It is sometimes confusing to read descriptions of acidity since the same author may state that *acidity is lower*, in describing a particular region or solution, yet in the next paragraph may describe the same region or solution as having *a higher pH value*.

If we remember that a higher pH number (or value, or level or reading) is the same as a lower level of acidity (which is the same as being more alkaline) we will understand what is being said.

So pH7 is neutral and the lower the pH number (3 or 4, for example) the more acid the substance being measured. The higher the pH number (6 to 6.9, for example) the less acid or more alkaline it is, with anything above 7 being alkaline.

amounts of bifidobacteria and bacteroides, as well as some enterobacteria and yeasts. In the region of the lower ileum the population density may reach between 100,000 and ten million organisms per millilitre of contents, not as high a density as exists lower in the GI tract but a very large population indeed. The major influences on the type and density of colonization of this region are:

- The degree of gastric acidity.
- The efficiency of intestinal motion (peristalsis) which is faster in the jejunum and upper ileum than the lower ileum, allowing the lower small intestine to contain a richer and denser bacterial population.
- The type of diet.
- The rate of stomach emptying.
- Exposure to microbial infection (much higher in the tropical regions for instance).

D. *The large intestine*, which runs from the ileum to the anus, is divided into the caecum, the colon and the rectum. Between the ileum and the caecum lies an important valve known as the ileocaecal sphincter.

It can take anywhere from 18 to 68 hours (or more) for food residues to pass through the entire large intestine, which is the greatest reservoir of micro-organisms in the body.

It has been estimated that in parts of the caecum and colon the concentration of bacterial flow reaches a density of between one hundred billion micro-organisms (10^{11}) and one thousand billion (10^{12}) micro-organisms per millilitre of contents, probably the maximum population density which it is possible to achieve.

Studying the variations in both type and identity of the inhabitants of the large intestine usually depends upon examination of faeces which have been found to contain virtually the same proportions of different bacteria, yeasts etc. as are found in the large intestine.

In adults the major colonists, in fact over 90 per cent of the micro-organisms in this region, are anaerobic (i.e. they do not require oxygen to survive), or what are called facultative bacteria (these are able to survive in either an oxygen rich or poor environment). Amongst the largest group of anaerobic and facultative inhabitants are bacteroides and bifidobacteria, followed (in terms of numbers) by eubacteria, peptostreptococci and ruminococci. A lesser (but vitally important) population exists in the large intestine of lactobacilli (which are usually aerobic, or oxygen dependent, organisms), as well as enterococci and coliforms, which comprise altogether between 1 per cent and 5 per cent of the flora of the region.

The balance of the inhabitants consists, when we are in good health, of small numbers of yeasts, proteus, staphylococci, clostridia and aerobic sporefomeres.

Viruses are also often found in the large intestine where they are usually transients. Those most often found are enteroviruses (echoviruses), which, when health is good and the friendly flora intact, normally clear from the region spontaneously in a matter of weeks without symptoms being observed.

Interaction between cultures

It is fascinating to realize that there exists a strong interaction between different groups of residents in the colon where, for example, some bacteria (lactobacilli, for instance) use up available oxygen and so create a perfect environment for the anaerobic inhabitants, which would die were the oxygen left unused. This also effectively prevents other less desirable bacteria which might require oxygen from staying in the area. Some bacteria (such as *E. coli* and certain strains of *Lactobacillus acidophilus*) are well armed and able to deal with their rivals by producing specific antibiotics, whereas others (notably the anaerobic groups) produce different weapon systems, such as short-chain fatty acid complexes, including acetic and butyric acid which inhibit their rivals' ability to thrive.

No room for new arrivals

Every suitable physical space in the GI tract is occupied and colonized by what may be termed indigenous micro-organisms. The body does not try to eliminate these, and there seems to be no immune response to them. They have effectively, through evolution, become a 'part' of the body's ecology. They are accepted and indeed depended upon for certain tasks. In contrast there are transient organisms, acquired from food, water or other regions of the GI tract, which are capable of colonizing vacated habitats; after antibiotic treatment, for example. This can severely disturb the ecosystem of the region. *Candida albicans*, a usually docile yeast, is one such potential hazard. Under normal healthy conditions, however, all the vacant space is occupied.

Let us now have a closer look at some of the major role-players in this complex, but usually friendly, colonization process.

The lactobacilli and bifidobacteria

The major characteristic of lactobacilli — and the quality which gives them their name — is their ability to produce lactic acid from carbohydrates and similar compounds. In the main, they are able to withstand a degree of acidity which usually destroys other bacteria (*New Gould Medical Dictionary*, Blakiston 1970) which is only to be expected since they manufacture so much acid themselves.

This ability allows them to colonize areas which are forbidden to many other bacteria, giving them the power to create a hostile environment for their rivals, and thus enhancing their ability to colonize their particular niches in the GI tract.

Amongst the more important of this large group of bacteria are:

- *Bifidobacteria*. Not one of the genus lactobacilli, although it used to be considered so. These are the major inhabitants of the GI tract, and we will consider them separately from the lactobacilli, later in this section.
- *Lactobacillus acidophilus*. The most prominent friendly resident of the small intestine.
- *Lactobacillus bulgaricus*. A transient, i.e. non-resident, yet important bacteria in human ecology, as we shall see.
- *Lactobacillus casei*. A different species of transient lactobacilli, strain Shirota, used in making cultured milk 'Yakult' in Japan.

Let us remind ourselves of the list of requirements demanded of micro-organisms in order that they be considered useful as dietary or supplemental aids to health:

1. They should be a normal inhabitant of (or regular and useful 'visitor' to) the GI tract.
2. They should be capable of surviving the digestive processes.
3. They should be capable of surviving and growing in the intestinal regions, in the presence of bile acids, for example, and being able to colonize and attach to the surfaces of the gut.
4. They should be shown to produce beneficial effects.

5. They should remain viable and active before and after consumption.

The major organisms which we will examine in this section fulfil all of these requirements (*L. bulgaricus* do not colonize, but help others to do so). The major beneficial functions of lactobacilli (excluding bifidobacteria, which are described below) when taken as dietary adjuncts are:

1. They enhance and allow digestion of lactose (milk sugar) by producing the enzyme lactase, and in general exert one of their most useful functions for us — that of aiding in the digestion and absorption of nutrients.
2. Some strains act to destroy hostile invading bacteria by producing natural antibiotic substances.
3. They have a direct anti-tumour effect.
4. They help prevent cancer by detoxifying or preventing the formation of chemicals which are carcinogenic.
5. They reduce the level of cholesterol, thus lessening the dangers of cardiovascular disease.
6. They help to produce important B-vitamins.
7. They are able to control proliferation of hostile yeasts such as *Candida albicans*.
8. Through their production of lactic acid they preserve and enhance the digestibility of foods which are fermented with them, such as soy products (tofu, miso), sauerkraut, pickles etc.

If these functions of the lactobacilli are disturbed the consequences can be severe in terms of our health. Studies have shown that apart from such obvious dangers as the use of antibiotics, stress can also severely affect the lactobacilli, reducing their relative numbers in the GI tract, altering the acidity of the region and producing changes in the availability of foods and vitamins usually absorbed by us with their symbiotic assistance.

The use of supplemental lactobacilli, in food or concentrated forms, can reverse such negative processes, and forms the basis of probiotics. When *Lactobacillus acidophilus* is supplemented, no adverse reactions are usually observed, unless the individual so supplemented is allergic to the base on which the acidophilus has been cultured (e.g. milk). In such instances, dairy-free cultures are available for which other growing mediums have been used (carrots, rice, soy, etc.).

Variations in benefits from lactobacilli

Not all strains of lactobacilli produce the same benefits, of course, since they have variable characteristics. Not even all strains of *Lactobacillus acidophilus* have the same potential benefits to offer.

In the previous chapter, we used the analogy of different breeds of dogs to illustrate the differences between different strains of lactobacilli. We are all aware of the difference between Great Danes and Dachshunds (*Lactobacillus bulgaricus* and *acidophilus* perhaps?) However, we should also be aware that within the dachshund breed there exist further variations — long-haired, short-haired, wire-haired, for example.

Thus it has been found that some strains of *Lactobacillus acidophilus*, such as DDS 1, can produce powerful antibiotic substances such as acidophilin, lactolin and acidolin, whereas other strains cannot. These natural antibiotics are produced by the acidophilus for its own benefit, but help to defend us when we are infected with pathogens, and they are extremely powerful.

In one study the acidophilin antibiotic produced by acidophilus strain DDS 1, when grown in a milk medium, was shown to strongly inhibit disease-causing micro-organisms such as *Salmonella typhimurium, Salmonella typhosa* (typhoid), *Salmonella schottmuelleri* (para-typhoid B), *Staphylococcus aureus* (toxic shock), *Vibrio comma* (Asiatic cholera) and *Clostridium perfringens*.

The lactobacilli also produce copious amounts of lactic, formic and acetic acids. These strongly inhibit other bacteria and micro-organisms which are not able to thrive in an acid environment. One such potentially dangerous inhabitant of the bowel is *E.coli*, which these acids effectively inhibit. Tests show that at a pH of 4.5 (acidic) lactobacilli thrive, whereas many other bacteria, including pathogenic (disease causing) ones such as streptococci and micrococcus die.

Lactobacilli use oxygen themselves, and also reduce the local oxygen supply through their production of various metabolites. This makes the survival of other competing oxygen-dependent micro-organisms less likely. A final control factor which they exert over enemies (both theirs and ours) is to very effectively compete for available food.

Why lactobacilli are so effectively supplemented

The fact that lactobacilli enjoy an acid medium helps them survive intact, whether taken in food or supplementally, on their way through the stomach with its digestive acids and the small intestine with its bile acids.

It would be relatively useless for them to survive the digestive process if they were then unable to attach themselves to the surfaces of the lower small intestine or the large intestine. This they do very well indeed — many studies show the ability of some strains to implant and adhere to be excellent. This requires of course that there be available space for such adherence, there being more chance of this after a period of antibiotic treatment or during, and immediately after, an episode of acute diarrhoea, for example.

Naturally enough, the higher the concentration of live, viable cells which are being supplemented, the better the potential for results in terms of colonization and subsequent health benefits. This would seem to be best achieved using concentrated cultures (powdered or encapsulated, with guaranteed potency and expiry date) rather than fermented dairy foods (yogurt, acidophilus milk, etc.) which are better used in general dietary situations rather than therapeutic ones.

The transient friendly bacteria

When milk is turned into yogurt two cultures can be used, these being *Lactobacillus bulgaricus* and *Streptococcus thermophilus*. Although not normal residents of the human GI tract, some strains of these bacteria exert very strong benefits when eaten or, more usually, when taken supplementally, not least a definite enhancement of the health and vigour of the bifidobacteria (see below) the most populous group of indigenous friendly resident bacteria in healthy people. Unfortunately, commercial yogurt bacteria tends to use less hardy strains (in terms of surviving digestion) than those employed in supplementation.

L. bulgaricus and *S. thermophilus* have been shown to produce substances with powerful anti-tumour activity. Another study, which will be discussed later, shows *L. bulgaricus* to have powerful anti-herpes virus effects. As with other lactic acid-producing bacteria, these transient bacteria encourage a more acid environment in which there is strong inhibition of less desirable micro-organisms which might be attempting to colonize regions of the GI tract.

Other benefits to us include enhanced digestion of the products in which they are found, such as milk solids; reduced chances of infection; and reduced likelihood of cancer-causing substances being formed in the colon. Some researchers believe that it is more beneficial to supplement with these transient bacteria than with the actual resident cultures such as acidophilus and bifidobacteria because they so obviously improve the health of such permanent colonies as they pass through.

Lactobacillus bulgaricus has sometimes been found to appear in faeces up to 8 days after supplementation had begun, and to continue being excreted for up to 12 days after supplementation had ceased, indicating that although a transient, it certainly stayed around long enough to have an effect on the ecology of the region.

It might be sensible, therefore, to supplement all the friendly bacteria when circumstances demand.

Bifidobacteria

The main friendly occupant of the large intestine in health adults, and in breast-fed infants, is a group (genus) known as the bifidobacteria. In infancy, the type of food received seems to determine the balance between the highly desirable bifidobacteria and less desirable putrefactive micro-organisms such as the bacteriodes.

Breast-fed babies, and adults on a pattern of diet which enhances bifidobacteria, have high levels of these. The presence of large colonies (up to 99 per cent of the total bacterial count) of bifidobacteria in breast-fed infants provides them with a greater protection against infection by keeping the GI tract relatively acid. Thus, a breast-fed baby will have a faecal acidity of around 5.0 compared with an almost neutral level of between 6.0 and 7.0 in formula-fed babies (and most adults).

The levels of bifidobacteria seem to decline with age and also in conditions of chronic ill-health: it is the opinion of many that this decline may precede and actually encourage the chronic changes seen in unwell elderly people.

Bifidogenic diet

A high complex-carbohydrate diet (vegetables, whole grains, pulses, fruits etc.) encourages high levels of bifidobacteria in adult life, and such a diet is therefore termed bifidogenic. A meat-rich diet has the opposite effect, encouraging putrefactive bacteria (e.g. bacteroides) to flourish. Putrefaction of food by such bacteroides is accompanied by a host of undesirable chemical substances being produced, such as ammonia, phenols, indole, amines and faecal bile acids. All of these are potentially harmful and are often associated by researchers with the high incidence of bowel cancer in societies that are heavily meat eating. Other bifidogenic factors (things that are helpful to the condition of bifidobacteria colonies) include the regular intake of helpful transient bacteria, such as *Lactobacillus bulgaricus* and *Streptococcus thermophilus*, although just how they achieve their benefits on the bifidobacteria is not yet clear. Swedish research seems to indicate that the assistance comes from both the lactic and acid which they produce and from the breakdown products which are derived from these transient bacteria themselves, and which provide acceptable 'food' for the bifidobacteria.

The role of bifidobacteria

1. The prevention of the colonization of the intestine by invading hostile pathogenic bacteria or yeasts with which they compete for attachment sites and nutrients.
2. The production of acetic and lactic acids, which lower the pH (increase the acidity) of the intestine, thus making the region undesirable for other, possibly harmful bacteria.
3. The retention of nitrogen, and assisting of weight-gain in infants. (Bifidobacteria have actually been called 'the key to infant nutrition'.)
4. The inhibition of bacteria which can alter nitrates in the bowel (derived from food and water) into potentially harmful nitrites, which are known to be toxic with a potential for causing cancer.
5. The manufacture of essential B-vitamins for our use. A healthy

bowel flora in both infancy and adult life obviously offers major health and well being benefits.

It can become disturbed due to all or any of the following during infancy:

- Sudden changes in nutrition
- Use of antibiotics
- Common infections
- Vaccination
- Convalescence from illness
- Sudden weather changes

In adults a more drastic reason is usually required to alter the bacterial flora balance, such as:

- Use of antibiotics and other drugs (steroids for example)
- Disturbed gastric function
- Disorders of intestinal motility (constipation, spastic colitis, diarrhoea etc.)
- Altered acid production from the stomach (production commonly lowers as we age)
- Conditions such as pernicious anaemia (which is accompanied by altered stomach acidity)
- Stagnation of intestinal contents (chronic constipation etc.)
- Changes in the structure or functon of the GI tract such as diverticulosis or regional enteritis (Crohn's disease)
- X-ray irradiation or other radiation exposure
- Cirrhosis of the liver
- Immune system deficiency
- Most chronic disease states are accompanied by (many say they are preceded by) altered GI tract flora

If for any reason stomach acidity levels decrease markedly, there is a tendency for overgrowth of various species of micro-organism which may then spread into the stomach and upper regions of the small intestine. This is most common in conditions of hypochlorhydria (absence of, or very low levels of, hydrochloric acid) where levels of stomach micro-organisms have been seen to rise to between 100,000 and 10 million per millilitre of stomach surface content, as compared with a usual level of around one thousand. When environmental conditions change in this way (i.e. less acidity) there is a tendency for bifidobacteria to become reduced, and an increase to be seen in potentially harmful inhabitants such as staphylococci, proteus, clostridia, klebsillia-enterobacter groups and, of course, yeasts such as *Candida albicans*.

Bacterial overgrowth by less desirable organisms in these regions may have very harmful consequences, resulting in a variety of possible conditions including megaloblastic anaemia, fat malabsorption etc. The re-introduction of lactobacilli and bifidobacteria, in the diet or supplementally, as well as removal of the causes of the original disturbance usually corrects such problems.

When do the friendly bacteria arrive and take up residence?

Before birth our gastro-intestinal tract is sterile, free of bacterial colonies, but during the birth process and immediately afterwards we acquire from our mothers, and from the environment, a host of bacterial 'contamination', including enterococci, lactobacilli, coliforms and other micro-organisms. These rapidly colonize the digestive tract. If a baby is breast-fed, bifidobacteria start to appear in the stools anywhere from 2 to 5 days after birth, and by the end of the first week of life the bifidobacteria form anything from 85 to 99 per cent of the colonizing bacteria found in an infant's stools.

We will see in Chapter 13, which deals specifically with children's needs for lactobacilli and bifidobacterial supplementation, that dramatic changes have been observed in the number and type of bifidobacteria found in infant GI tracts over the past 30 years, with environmental contamination of air, food and water involving radioactivity, antibiotics, pesticides and other changes apparently leading to a major shift in the internal ecology of modern children.

The pattern of bifidobacteria colonization is less strongly observed in premature infants and in formula- (or cows-milk) fed infants, with other micro-organisms such as bacteroides and various lactobacilli being present in larger numbers. A decline in the strain *B. infantis* is concerned (states Dr J. Rašić, a world expert on fermented milk products) with negative environmental conditions.

A change also takes place in the intestinal flora balance on weaning, although if breast milk continues to make up 50 per cent or more of the diet during weaning the levels of bifidobacteria colonies in the GI tract stay relatively high.

As we grow older, the relative levels of these bacterial groups alter. Clostridia, enterobacteria and streptococci increase as the production of gastric acids gradually reduces with age, lessening the acidity levels in the regions of the GI tract. There are now thought to be 24 species (and numerous sub-groups) of bifidobacteria, each with slightly different physical characteristics, physiological behaviour and patterns of fermentation.

All bifidobacteria occurring in man have the ability to ferment lactose (from milk), as well as glucose (sugar) and fructose (fruit sugar). Some strains are also able to ferment other nutrient breakdown products, thus enhancing the digestibility of these foods.

At different periods of life, and under different conditions, the relative presence of species of bifidobacteria varies considerably. There is no general agreement about which species of bifidobacteria predominate in breast-fed and formula-fed infants. Earlier reports showed that *B. infantis* was dominant in breast-fed babies, but some subsequent research showed *B. bifidum* to be more common, with *B. longum* dominating in formula-fed babies. Dr Rašić has shown that in modern breast-fed infants, from 1980 onwards, *B. infantis* seems to be in decline, with *B. bifidum* now dominating. Other research published in 1984 found *B. infantis* in more babies than other bifidobacteria types. Recent Japanese research differs and shows *B. breve* to dominate in both breast- and bottle-fed infants. According to Dr Rašić, 'the possible explanation for the above changes may be . . . a consequence of an increasing contamination of the human environment . . .'

The two species *B. adolescentis* and *B. longum*, commonly found in bottle-fed infants, are also the major organisms of the friendly bifido genus present in adult and the elderly.

As we will see in later sections, which deal with specific health problems, we have in the bifidobacteria a strong and useful ally which can be used therapeutically and preventively in a host of disease states, with no side-effects normally experienced. We need to establish, of course, the ideal way in which this assistance can best be achieved, and also to show that supplementation, in either food or concentrated form, is effective. This will be done in the coming chapters.

A different opinion

Some experts argue that the likelihood of supplemented acidophilus actually finding a niche in which to adhere, grow and flourish is minimal, except after a recent course of antibiotics. Rather, it is suggested that in appropriate states of ill-health, there should be supplementation of some of the beneficial yet transient bacterial strains. One example is *Lactobacillus bulgaricus*, which acts as a stimulus for the resurgence of bifidobacteria activity when this has been negatively affected for any reason.

This argument is put forward by Scandinavian researchers but must remain a theory until stronger evidence is provided. There is, however, no reason why the best of both concepts should not be adopted: in suitable cases, both acidophilus and bifidobacteria are supplemented along with transient cultures such as *streptococcus thermophilus* and *Lactobacillus bulgaricus*, the main cultures used in the making of real yogurt.

Chapter 3

How the friendly bacteria help your nutrition

Early this century Dr E. Metchnikoff wrote a book entitled *The Prolongation of Life* (Putnam, New York, 1908) in which he expressed the view that the extreme longevity observed amongst the people of Bulgaria probably related to their huge consumption of fermented milk products. Metchnikoff specially credited yogurt, and its major bacterial culture *Lactobacillus bulgaricus*, with this health benefit. He considered that 'poisons' with undesirable effects on health were often being produced by unhealthy bowel flora and that those cultures present in fermented milk products could balance and neutralize these toxins.

He was not very far from the truth, as modern research has shown. Naturally occurring food fermentation processes induced by lactic acid bacteria are amongst the oldest forms of food processing and preservation known to man, dating back into prehistory when organisms from the environment (in the atmosphere, present in or on foods, or utensils) performed this task unasked and unaided.

Drs Rao and Shahani say that: 'According to legend, the method of preparing yogurt was revealed to Abraham by an angel, and he owed his fecundity and longevity to this fermented food. Vedic literature (6000 BC) from India associates longevity to the consumption of dairy products.'[19] (The earliest evidence of the domestication of cows dates to 9000 BC in Libya.)

With time people learned to enhance the natural, but apparently unpredictable, process of fermentation by adding specific cultures and providing an ideal environment in which 'souring' could take place, including elements such as temperature control, the level of acidity, the time allowed for fermentation, the water content of the food being so processed, and so on.

Foods which were traditionally preserved, enhanced and processed in this way include yogurt, acidophilus milk, koumis, cottage cheese, kefir, cheese, sausages, pickle, sauerkraut, sourdough breads, oriental foods (such as tofu which is also known as soya cheese or bean-curd, tempeh, miso etc.) kvass and many more.

Indeed there is hardly any society which has not since ancient times used fermentation, as evidenced by the more than 700 varieties of cheese which are produced from the milk of almost every conceivable animal ranging from camels to yaks. It is generally agreed that, as in so many things, the Chinese were probably the first to use these methods for preserving vegetables (pickles).

Among the benefits of such processes are prolonged storage life; avoidance of the need for further heating or cooking, which protects the nutritional value as well as creating a degree of stability; and allowing the vitamin content to be retained for longer periods, during storage.

What makes a food 'good' for us

For a food to be of value to the individual who eats it there should be not only a reasonable food value present, in terms of its content of one or more of the 40 odd nutrients (essential fatty acids, minerals, trace elements, vitamins, enzymes, protein, carbohydrate, etc.) which we require for survival in good health, but there should also be evidence of such nutrients being what is termed 'bio-available'. This means that the food should be digestible, and the nutrients relatively easy to assimilate by the person eating it.

One of the major benefits bestowed by the various fermentation processes involving lactic acid bacteria is the enhanced digestibility of the food so treated. Indeed, in many instances, the food is virtually predigested by the friendly fermentation bacteria.

Protein and fat bio-availability

Not only are nutrients such as protein and fats made more bio-available when fermented, in yogurt for example, but the protein level is often increased mainly by fortifying milk with non-fat dry milk or by concentrating the milk, and to some extent by the fermentation process.

Protein, as the human body uses, it, needs to be in the form of either free amino acids (the individual building blocks which constitute protein) or in very small amino acid chains (peptides). This is what our digestive enzymes and acids have to accomplish in order to make any protein we may eat of any use to the body to meet its multiple requirements. It is in just this form that protein is presented to our digestive systems when natural yogurt, containing lactobacilli such as *L. bulgaricus*, is consumed. The enzymes from these lactobacilli begin the process of digestion of proteins and fats, as well as of milk sugars (lactose), even before we eat the food.

Shahani and Friend, in their studies of this subject have shown that there is up to a 10 per cent improvement in the quality of the protein (calculated on specific chemical and biological measurements) found in cultured yogurt, compared with the skimmed milk from which it was made.

Rašić has compared different protein-rich foods and gives their 'biological' value as follows:

Food	Biological value
Maize (corn)	59.4
Whole wheat	64.7
Chicken (cooked)	74.3
Beef (cooked)	74.3
Fish	76.0
Cow's milk	84.5
Yogurt	87.3
Yogurt (sheep)	89.3
Yogurt (goats)	90.5

When animals are fed on yogurt, for example, the efficiency with which they digest and utilize the food is indicated by their rate of growth compared with animals fed on other food sources such as those listed. This is one way in which the biological value of food can be measured.

The protein found in yogurt is far less likely to provoke allergic reactions in people who are sensitive to milk, since the effect, of lactic acid on the protein will have made it more bio-available, in smaller fractions (peptides and free amino acids).

The process of fermentation increases the protein level of the food being cultured, so that fermented protein foods actually contain more protein after the process of fermentation than they did before the culture was added.

It seems that when *Lactobacillus bulgaricus* (as an example) is used to culture cheese whey into yogurt, there is a net gain of 7 per cent in the protein content of the final product, this additional protein being directly derived from the lactobacilli cells.

This means that we not only receive partially predigested protein, in a more bio-available form, but also a bonus of additional protein when such foods are consumed (if, that is, they actually contain the appropriate bacterial cultures, for unfortunately in modern commercial processing this is not always the case, as will be explained below). These benefits have also been proved in relation to the digestion of beans and their flours, especially in the Far East where fermented forms of these plant foods comprise a major element of the diet.

In many of these plant based foods there are found substances which inhibit their easy digestibility, including chemicals known as phytates and trypsin-inhibitors. These anti-nutritional elements are absent when foods such as soya have been fermented, indicating that the chemicals are either destroyed by fermentation or are prevented from operating.

One other not insignificant benefit found in relation to the use of the fermented versions of such foods is the removal of the 'flatulence factor', which produces so much distress for some individuals. The presence of streptococci such as *S. thermophilus* (see previous chapter) ensures that the fat content of the milk is also partially predigested. Studies have shown that when fermented foods are consumed they retain a far higher nutritional value than their unfermented raw materials (milk etc.). This can be translated into improved physical performance in, for example, animals fed on cultured products as opposed to those fed on the raw unfermented product such as milk and soya beans.

Enhancing digestion of milk products

More than half the population of the world has inadequate levels of the digestive enzyme lactase in their GI tract. Lactase has the job of converting lactose or milk sugar into a form the body can use such as glucose or galactose. This missing enzyme makes the digestion of milk and its products difficult, and sometimes impossible, for such people. Lack of lactase is more prevalent in people of Southern European, African and Asian origin, with Northern Europeans being singularly more able to digest milk in all its forms.

When lactose (milk sugar) is indigestible, because of lactose deficiency, a number of intestinal problems arise if the person consumes milk or dairy products. Such people understandably and necessarily should restrict their intake of these foods.

Some authorities argue that there are other and better ways than the consuming of milk to gain access to nutrients such as calcium, and it is well established that if the diet is wholesome and balanced, milk itself is not an essential food anyway. However, there is an undoubted value in people being able to digest milk, especially where protein deficiency is common. One of the major benefits derived from active lactobacilli such as *L. acidophilus* and *L. bulgaricus* is that they supply the enzyme lactase, which then enables the consumer to digest milk and to gain access to both the calcium bound in it and its abundant protein.

In one study it was demonstrated that people who were lactose intolerant, and who were, therefore, unable to digest skimmed milk, had the ability to digest yogurt (because its cultures were producing lactase) with the same degree of efficiency as people, who possessed adequate lactase of their own, were able to digest skimmed milk.

Similarly, people who reported abdominal problems (distension, diarrhoea etc.) when given milk, reported no symptoms when given yogurt. These benefits are not found in pasturized yogurt which has no active cultures.

> Note: We must differentiate between yogurt made with milk (or skimmed milk) which has been pasteurized and yogurt which is pasteurized after its cultivation.

Heat treatment of milk *before* it is cultured is vital in order to obtain vigorous growth of the culture and to get what is known as a good 'set'. If unpasteurized milk is used for yogurt, an uneven end-product emerges with poor growth rate for the friendly micro-organisms. The beneficial nutrients destroyed during pasteurization of milk are more than compensated for by the new nutritional benefits guaranteed by the culturing process.

Similar benefits can, of course, be gained by using supplemented (powdered for preference) cultures of active *L. acidophilus* or *L. bulgaricus*. When yogurt is pasteurized *after* the culturing process, this destroys some of the nutrients, and almost all of the friendly bacterial activity.

Calcium levels of different foods

Eight ounces (225g) of each of the following foods contains the indicated level of calcium (in milligrams). The second figure given indicates what percentage of the recommended daily allowance this amount of calcium represents, for adults.

Whole wheat bread	190mg	24%
Potato	13mg	2%
Chicken	18mg	2%
Beef (off the bone)	27mg	3%
Fish (canned tuna)	36mg	5%
Whole milk	291mg	36%
Low fat yogurt	415mg	52%

Professor Rǎsić points out that calcium and other minerals are better absorbed from yogurt than from milk because the combination of acid and the improved protein digestibility greatly enhances calcium absorption. In the next chapter, which deals with damage to the internal ecosystem of the GI tract, we will discover that alterations in the health of the flora of this region can have a most surprising effect on the circulating levels of female sex hormones such as oestrogen. A consequence of this on calcium metabolism is explained in the section 'Two unexpected consequences' on page 67.

Manufacturing B vitamins for us

The benefits which the lactobacilli confer on us are not entirely altruistic. They do not produce antibiotics for our benefit, but to protect themselves against organisms which are harmful both to themselves and to us. We benefit from this, of course, but this is not the intention of the friendly bacteria, for they have no intent apart from the survival of their species. Similarly, when we observe that there are many instances in which the addition of micro-organisms to a food, for purpose of fermentation, produces a rise in that food's level of certain B vitamins, we can be sure that this bounty was produced not for us but for the bacteria's own purposes.

The fact is that many lactobacillus cultures require additional levels of the B vitamins in order to grow, and their manufacture of these, from which process we benefit enormously, relates to their needs and not ours. Doubtless, were there not such benefits accruing to us, from their presence inside us, our bodies would long ago have found ways of getting rid of them. As it is, we can now hardly survive without them.

Professor Rašić says: 'Changes in the vitamin content occurring during the manufacture of cultured milks are dependent upon many factors. Some bacterial strains synthesize vitamins during growth, whilst others consume them. For instance, streptococci and yogurt bacteria (bulgaricus) synthesize folic acid, while acidophilus bacteria consume it. Some yogurts and acidophilus milks are richer in B_{12} and others poorer. ('Role of Dairy foods containing bifido and acidophilus bacteria in nutrition and health.' *North European Dairy Journal* 4/83 pp 80–88.)

Those B-vitamins which are most markedly enhanced by certain of the lactobacilli are:

Biotin	Increases in cultured yogurt
Vitamin B_6 (pyridoxine)	Increases in cheddar cheese, cottage cheese
Vitamin B_{12} (cyanocobalamine)	Increases in cottage cheese
Folic acid	Increases in buttermilk, sour cream, yogurt, bifidus milk, kefir
Vitamin B_3 (niacin)	Increases in yogurt, cottage and cheddar cheese
Vitamin B_2 (riboflavin)	Increases in yogurt
Vitamin B_5 (pantothenic acid)	Increases in cheddar cheese
Vitamin K (the anticlotting factor)	Manufactured in the GI tract by certain strains of friendly bacteria

(adapted from Shahani and Chandan 1979)

Note: These increased levels apply only to cultured products and not those produced by direct acidification processes.

Examples of the way B vitamin levels rise in certain foods include:

- Cottage cheese which has a content of up to 5 micrograms of folic acid per 100 grams, compared with less than 1 microgram per 100 grams being present in the milk from which it is cultured.
- Cheddar cheese which has a content of up to three times the B_6 found in the milk from which it is cultured (rising up to 147 micrograms).
- Yogurt which has a content of 3.9 micrograms of folic acid per 100 grams compared with the starter milk's level of between 0.13 and 0.73 micrograms. Yogurt bacteria also often double the levels of B_3 (rising from between 71 and 96 micrograms in the starter milk to between 130 and 141 micrograms per 100 grams in the yogurt).

Dr Shahani also shows that Vitamin A levels rise from a level of 9 micrograms in skimmed milk (per 100 grams) to between 70 and 130 micrograms in plain low-fat yogurt.

Problems due to production methods?

Dr Khem Shahani, of the University of Nebraska, and Dr Ramesh Chandan, of Michigan State University, writing in *The Journal of Dairy Science* in 1979, described one of the problems which we are bound to find in relation to modern dairy products. For, while it is clear from the brief examples given above (in the instance of cottage cheese for example) that there are very definite nutritional advantages to be gained by consuming such fermented foods, this is not by any means a certainty.

As Dr Shahani and Dr Chandan state: 'Newer processes have been developed to manufacture yogurt, cottage cheese, buttermilk and sour cream by direct acidification instead of by lactic cultures. The advantages of such processes over the conventional culturing methods are (a) elimination of culture handling problems; and (b) better quality control and uniformity of production; and (c) better keeping quality.'

These researchers make it clear that these undoubted commercial benefits are at the expense of nutritional value. This becomes clear when we are told that instead of folic acid rising to levels of up to 5 micrograms per 100 grams in cultured cottage cheese, the level reached when this is produced by acidification is only 0.1 microgram!

Not only nutritional quality is lost, they inform us, for: 'They (acidified products) are normally more bland, and lack the characteristic pleasant flavour and aroma commonly associated with cultured products.' A search of health food shops and specialist dairy suppliers should, nevertheless, still enable purchasers to track down foods prepared with traditional cultures.

The nutritional benefits of fermentation are not confined to dairy foods, for in certain wheat fermentation processes (e.g. idli and khaman) there is found to be a dramatic rise in Vitamins B_1 (thiamine) and B_2 (riboflavin). This is particularly important in the regions of the world where malnutrition is common.

Other nutritional benefits

A problem which periodically recurs, associated with processed meat products such as sausages, is food poisoning, the result of contamination by certain staphylococcal bacteria which produce toxins. When lactic acid fermentation cultures are used in the preservation of sausages, however, this is not possible, as the enterotoxins are not produced by the staphylococci in the presence of lactobacilli.

A similar protection is exerted over the formation of toxins by the infamous food poisoning agent *Clostridium botulinum* when lactobacillus cultures are used in combination with suitable sugars in preserved foods. Specific benefits were demonstrated in studies of people infected by foodborne pathogens such as *Escherichia coli* and *Salmonella typhimurium* when they took *L. acidophilus*. This effectively suppressed the growth of these micro-organisms. We will examine this protective phenomenon more closely when we deal with the antibiotic qualities of some of the lactobacilli.

As we ponder on these multiple benefits we should keep in mind the possibility, touched on briefly in this chapter, that commercial considerations may deny us access to the ideal versions of many of these foods. This very real problem will be dealt with in greater detail in Chapter 15 which discusses commercial influences on the foods discussed above and on the different versions of these marvellous bacterial cultures which can be used supplementally for so many health problems.

The next chapter will examine how a healthy environment in the GI tract can be disturbed, as well as the consequences of such changes, and what we can do about it.

Chapter 4

How and why the ecosystem in the gastro-intestinal tract changes

The health and integrity of the multiple populations which make up the normal bowel flora can be affected for good or ill by a variety of internal and external elements, and any such changes can in turn make a dramatic impact on the overall level of health (or disease) we experience. Disturbed ecology of the intestinal flora can involve exessive levels (overgrowth) of undesirable micro-organisms, or very sparse levels of the more desirable residents, or patterns of colonization in regions which remain relatively uncolonized when we are in good health.

In ideal laboratory conditions bacteria can divide every 20 minutes to form two new micro-organisms. Thankfully such ideal conditions seldom prevail inside any of us, or we would quite simply be overwhelmed by the sheer volume of these tiny creatures. In fact, researchers show that it is probable that each bacteria inside our GI tract divides no more frequently than four times per day, and possibly only once. This retarded rate (as compared with what is known to be possible) seems to be directly connected to a number of elements, such as those discussed below. These are also the major factors impacting on the overall well-being of the friendly flora.

Internally arising change factors

1. *Normal peristaltic action*. The rhythmical contraction of the small intestine prevents colonies from setting up residence there, for when this function is slowed down a rapid overgrowth of this region takes place.

2. *Stomach acid levels* keep the population of the stomach and

small intestine down, and when this acid production is diminished, for whatever reason, this can lead to massive overgrowth and the presence of undesirable elements (which adequate acid levels keep at bay) such as yeast colonies. One consequence of frequent use of antacids, such as bicarbonate of soda, is the fact that relatively fewer harmful bacteria are required to enter the system before infection results.

Other secretions in these regions, such as pancreatic enzymes and bile acids, have not been shown to have a marked effect on bacterial growth.

3. *The interaction of one type of bacteria with another* is very important in determining the pattern of colonization. Unless oxygen dependent bacteria are present in sufficient numbers to reduce oxygen levels, those which require an oxygen-free environment, such as the bifidobacteria, cannot survive in abundant good health. There are also the antibiotic secretions of certain species (*E. coli* for example) and the fatty acid secretions of others (bifidobacteria) which regulate the presence of their rivals, as do the acid secretions of some of the friendly bacteria, which create barriers to other cultures who require a more alkaline environment.

Externally arising change factors (and consequences of such change

4. *Antibiotics*. Almost any antibiotic administration by mouth can severely alter the balance and pattern of the intestinal flora. This will vary from one medication to another depending upon the spectrum of the drug (the range of organisms which are sensitive to it) and also to the concentration of the antibiotic when it reaches the bowel.

The reduction in total populations of the flora caused by antibiotics can be dramatic, but is usually short-lived. This fact is used in treating people before abdominal surgery in order to minimize the likelihood of post-operative sepsis. Studies at the Veterans Administration Hospitals in the USA show that use of drugs such as erythromycin and neomycin, before such surgery, results in both aerobic and anaerobic bacterial levels dropping to between 20 and 25 per cent of the original levels. This leads to a reduction in post-operative wound infection — only 9 per cent of patients contract infections compared with 35 per cent of those not receiving antibiotics of this type.

That is the positive side of the picture. Unfortunately, the ecological vacuum created by the indiscriminate slaughter of friendly bacteria is often rapidly filled by potentially pathogenic (disease causing) micro-organisms.

Examples of such rapid repopulation are numerous and include an overgrowth syndrome called *Pseudomembranous enterocolitis* caused by the organism *Clostridium dificile*, a notorious opportunist which populates the large intestine during treatment with antibiotics, especially ampicillin. The effect of the toxins produced by this unpleasant interloper is to destroy the surface of the colon, producing ulceration and consequent symptoms of bloody diarrhoea, pain, and catastrophic weight loss etc.

Similar symptoms result when the population of another potential pathogen, *Staphylococcus aureus*, explodes and produces its own version of enterocolitis, after the use of broad-spectrum antibiotics.

A phenomenon known as *super-infection* may follow the use of antibiotics. First, the strains of friendly flora which are sensitive to the antibiotic die and there is a proliferation into this vacuum of micro-organisms which are resistant to the drug. These may be bacteria or yeasts already living in the bowel, albeit in controlled quantities, or they may be newcomers implanted from the environment.

The micro-organisms most involved in super-infection, which can occur in almost any region or organ of the body, following antibiotic treatment, include Staphylococcus, Pseudomonas and Proteus species; Influenza bacillus, and *Candida albicans*.

Candida albicans in particular deserves mention since a wide range of symptoms are possible as a result of its rapid overgrowth following antibiotic treatment. These include: inflammation of the tongue (glossitis) and mouth (stomatitis), vaginitis and proctitis (inflammation of the rectum) as well as a range of mento-emotional symptoms such as anxiety, irritability and depression. Researchers such as Dr Orion Truss and Dr William Crooke in the USA have also shown that multiple allergies, digestive symptoms (bloating, heartburn, constipation, diarrhoea etc.), cystitis, premenstrual syndrome, fatigue, acne and other skin problems all may relate to yeast overgrowth.

> Note: see box on page 99 which shows how dramatically the presence of Candida can stimulate some bacterial infections, making them life-threatening.

Chapter 8 is largely devoted to the identification and treatment of yeast proliferation, since one of the most effective therapeutic weapons we have against Candida involves the supplementation of high potency cultures (of a guaranteed potency with an expiry date) which can control it.

5. *Sex-hormones* have been shown to have a most important relationship with the intestinal flora. It has been shown that in normal health over 60 per cent of circulating female hormones such as oestrogen are excreted into the GI tract in the bile. The hormones are then, in normal conditions, acted upon — a process known as deconjugation — by bacterial enzymes such as sulphatase catalyse before most of it (some is excreted in the faeces) is re-absorbed into the bloodstream. From there, it is sent back to the liver for reactivation into a biologically active form.

Some of the oestrogen is chemically changed on re-absorption by the mucosal cells of the bowel into a form which cannot be recycled via the liver, and it is then excreted in the urine. Since this process leading to urinary excretion can only take place in the lining of the colon, the level of oestrogen found in urine can be used to show the relative efficiency of these processes (re-absorption etc.) in the GI tract.

When antibiotics (ampicillin, penicillin, neomycin etc.) are in use the levels of oestrogen products found in the urine are very much lower than normal. The levels of re-absorbed oestrogen are also lower when antibiotics are used, and the quantity found in faeces is increased by up to 60 times the normal levels.

These changes naturally lead to lower levels than would normally be circulating (being present) in the bloodstream and are thought to result from the damage done to the microflora by antibiotics.

The effects on women can be unpleasant, with what is termed *breakthrough bleeding* being all too common in women when antibiotics are used. More seriously, there is a frequent incidence of oral contraceptives *failing to prevent pregnancy* at such times, since circulating levels of sex hormones would be reduced due to the effects of antibiotics on the bowel flora. Another consequence of this reduction in oestrogen levels would be expected to involve *osteoporosis*. This possibility is discussed further in the section 'Two Unexpected Consequences' following the next chapter.

6. *Stress* has been shown to induce changes in the balance of the flora, mainly resulting in reduction or disappearance of lactobacilli and bifidobacteria, a reminder that holistic thinking should take account of *all* influences on the body, not just the obvious ones.

7. *Radiation* can cause great damage to the flora, either when direct x-rays or gamma radiation is performed medically on the abdominal region or when radioactive materials are swallowed.

8. *Dietary changes* can influence the health and behaviour of intestinal flora although there is controversy in this area of study since the method used to measure changes in type and numbers of different micro-organisms depends upon examination of fresh faeces.

Drs Simon and Gorbach discuss this in their review of this subject (see page 216). After explaining how numerous studies indicate that dietary changes do not seem to significantly alter the major constituents of faecal flora, they say:

'It must be emphasized that these diet studies were performed with classic bacteriological techniques of counting and identifying micro-organisms in a measured amount of fresh faeces. *Such techniques are imprecise at best, in view of the reliance on cultivation methods, and are highly liable to error in a system of 100 billion bacteria, comprising at least 40 types of anaerobic bacteria*!'

Far more important in terms of the health of our intestinal flora and its relation to our eating habits is their next remark: 'Studies of metabolic activity of the flora based on measurements of bacterial enzymes *have in fact revealed marked changes in the colonic flora as a function of diet*.'

This statement cannot be emphasized too much, for it is often suggested that because the bacterial levels of the different inhabitants do not seem to change much in response to diet alone (at least not in short term studies anyway) there is no real need to concern ourselves with the diet as a factor in altered bowel ecology. The fact is that the inhabitants may not change much, but what they *do* alters enormously and with profound effects on our general health. There is also evidence that over time the actual populations begin to change, especially when we look at the very young digestive tract.

Examples of such influences include:

Weaning: It is clear from the process of change in the type of inhabitant of the GI tract after breast feeding ceases that diet strongly influences the flora, at this stage of life at least.

There is a gradual increase in numbers of putrefactive bacteroides over the months when breast fed infants are completely weaned until their numbers match those of the bifidobacteria, formerly the overwhelming occupants of the colon. At this time, other mixtures of bacterial flora appear in the colon including clostridium, escheria, streptococcus and lactobacillus, and these changes set the scene for diarrhoea.

Behavioural changes due to diet: Drs Fernandes, Shahani and Amer believe that dietary changes which alter the metabolic activity (not necessarily the numbers) of anaerobic bacteria, such as the bifidobacteria, may reduce their ability to produce volatile fatty acids with which they control less desirable (e.g. coliform) bacteria, allowing these to become more active or to proliferate and to thus produce symptoms.

Research into the way diet can influence the development of bowel cancer, and the ways in which the friendly bacteria beneficially influence matters by neutralizing harmful chemicals in the region — has resulted in our now having a better understanding of the way diet alters microbial behaviour. This is dealt with more fully in the next chapter.

It is clear that even if diet fails to alter the type and concentration of micro-organisms in the adult GI tract to any great extent, at least in the short term, it nevertheless changes dramatically the way these micro-organisms behave and the substances which they secrete.

One group of substances which can cause cancer is called polycyclic aromatic hydrocarbons, and these have a chemical similarity to the bile acids found in faeces.

If we eat a great deal of animal fat there is a corresponding increase in the levels of bile acids from the liver, and this in turn induces bacteria living in the colon to produce an enzyme called 7a-dehydroxylase, whose role is the breaking down of bile acids into what are called secondary bile acids.

People on a Western-style, high-protein diet (meat and therefore animal fat) have been shown to have in their colons more bacteria capable of producing 7a-dehydroxylase than do people living on foods of a more vegetarian nature, and people with colon cancer are found to have high levels of 7a-dehydroxylase as well as enzymes (derived from bacteria) which degrade cholesterol.

It is also found that populations who eat large amounts of meat contain in their bowels and faeces high levels of potential carcinogens such as coprostanol and coprostanone, which are breakdown products of cholesterol (high in meat eaters), whereas vegetarians from the same communities have low levels of these chemicals in their colons.

These findings have led to the conclusion that the bowel flora of meat eaters are 'more active metabolically' than are those in people on a diet which tends to more vegetables and complex carbohydrates.

When people change from a meat eating diet to a more vegetarian pattern, enzymes such as B-glucuronidase decrease in presence in their bowels. This is an enzyme which can actually retoxify substances which had been previously detoxified by the liver, changing them back into cancer-causing agents. Conversely, when a grain-rich diet is switched to a meat orientated one, this and other similar, potentially harmful, bacterial enzymes increase in activity.

One study showed that these negative changes can be induced by simply adding high levels of beef fat to a grain-based diet, even though no actual beef was included. The studies on which this evidence is based involve both humans and animals and can be regarded as conclusive.

Implications: The lesson we can learn from all this is that a potentially extremely hazardous alteration can be created in the secretions of the bacterial inhabitants of the bowel, although they themselves may remain the same in number, when dietary patterns change.

The findings of researchers who examined variations in bacterial flora status, as well as the secretions of the flora, in meat eaters, lactovegetarians and vegans, concluded that: 'Long term commitment to a vegetarian diet decreases the metabolic activity of the faecal microflora . . . in such a way as to probably reduce the likelihood of cancer of the colon'.

Another important research finding has been that when people on a meat and fat-rich diet are also fed supplemental *L. acidophilus* the undesirable enzyme presence is significantly reduced. In other words we may be able to reverse the negative trend induced by a poor diet by means of such supplementation.

This topic is dealt with in more depth in the chapter covering the ways in which the friendly bacteria protect us against cancer (Chapter 10). Its discussion here is to illustrate the way in which diet alters the behaviour pattern of these ubiquitous creatures.

9. *Acute diarrhoea* resulting from food poisoning (which we look at in more detail in Chapter 12) or infection, for example, will dramatically affect the friendly populations of the bowel. In fact at times the resident microflora will be completely overwhelmed by the activities of an invading pathogen (bacteria or virus) which produces diarrhoea via its toxic by-products.

The rapid bowel movements which occur in such circumstances lead to a marked reduction in the presence of bifidobacteria in the large intestine, although they usually rapidly recolonize it when the crisis is over. There is a difference between forms of diarrhoea in which the pathogenic bacterial cause is a colonizer of the tract (such as Salmonella, Shigella and *Escherichia coli*) and those in which the cause is a transient such as *Staphylococcus aureus*, some strains of *Clostridium perfringens* and *Bacillus cereus*. Obviously any bacteria which succeeds in colonizing the tract will be more difficult to control than a transient invader which would pass through naturally in time. The diarrhoea itself is the result of either the rapid proliferation of colonizing bacteria or of the toxins they produce.

Apart from adverse dietary conditions disturbing the micro-ecology of the bowel, it is common, especially in regions in which hygiene is less than stringent, for vast amounts of pathogenic bacteria to be consumed in food and water.

The excessive and often inappropriate use of antibiotics (such as in the case of virus infections over which antibiotics can have no influence) also sets the scene for diarrhoea by decimating the friendly bacteria. This is even more the case when broad spectrum antibiotics are used. It is estimated by Dr Shahani that half a million deaths of infants (children under age five) every year result from the diarrhoea caused by the proliferation of the pathogen *Clostridium fragilis* after the use of Clindamycin.

Whatever the bacterial cause of diarrhoea, as we shall see when we look at the antibiotic effects on the friendly bacteria (Chapters 5 and 7) we do have in many of these benign micro-organisms a weapon against this potentially fatal condition.

Polish studies show that the simple procedure of feeding acidophilus milk to children with Salmonella and Shigella dysenteries produced between 40 and 70 per cent faster recovery as compared with children not treated in this way. Japanese doctors gave acidophilus milk to patients with dysentery and achieved 100 per cent return to normal. Other excellent results that have been obtained in Japanese studies are discussed in the next chapter, which deals with the antibiotics produced by some of the friendly bacteria.

Whatever treatment is used in dealing with acute diarrhoea, it remains of paramount importance for the fluid and electrolyte balance of the individual with diarrhoea to be maintained, and while this may call for intravenous methods, it is usually possible to achieve satisfactory rehydration by drinking adequate amounts of liquid.

10. *Sexually transmitted infections* (including *Gay Bowel Syndrome*): Anal intercourse, whether indulged by homosexual males or in a heterosexual setting, can produce overwhelming damage to, and infection of the lower bowel. A condition originally called Gay Bowel Syndrome appeared in the mid-'70s, predating AIDS by a few years. It involved infection with a host of bacteria, viruses, protozoal micro-organisms, parasites and yeasts, transmitted to the lower bowel by sexual contact and resulting in chronic infection.

It should not be surprising that anal intercourse produces such a great degree of damage to the lower bowel. Not only is the region traumatized by this form of sexual behaviour, resulting in lacerations and abrasions and consequent ease of infection, but a number of other coinciding factors produce a lowering of local immune function. These include the introduction of sperm into the colon. Sperm is itself immune suppressive (it needs to be to allow fertilization to occur in vaginal sex — otherwise the sperm would be destroyed before this could happen).

The use of petrochemical-based lubricants are also known to be strongly immuno-suppressive, and these are usually employed in anal sex, due to lack of local lubrication.

The disturbance of the bowel's ecology in this manner has a marked effect in lowering local and general immune function, thus predisposing the individual to greater infection, nutritional deficiencies and a wide range of symptoms. Clearly this practice cannot be considered biologically normal. Any assistance which use of the friendly bacteria might offer (and this can be considerable) would be of short-term value if this particular practice were not abandoned. It appears that with the AIDS epidemic has come the practice of 'safe sex', leading to areas, such as San Francisco, now registering a dramatic decline in the incidence of new cases.

We will now move on to examine the amazing antibiotic properties of some of the friendly bacteria and how we can use these to our advantage.

Chapter 5

The antibiotic and antibacterial powers of the friendly bacteria in the gastro-intestinal tract

Of all the many benefits which we derive from their presence, none is more important than that of the amazing antibiotic, antibacterial, antiviral and antifungal potentials which some of the friendly bacteria possess.

Some specific strains of the lactobacilli produce powerful antibiotics, which actually kill other bacteria, for their own protection against rivals which are sometimes dangerous but always competitive. Others have as their means of protection a range of strategies which can inhibit the growth and activities of other micro-organisms without actually killing them as an antibiotic would do. These strategies include the production of acid and hydrogen peroxide (H_2O_2).

In this chapter we will present some of the evidence of this sort of protective activity by relating this to gastrointestinal problems and, in the following chapter, to vaginal and urinary tract disturbances. Among the most researched of these natural antibiotics we have:

Organism	*Antibiotic which it produces*
Streptococcus lactis	Nisin
Lactobacillus brevis	Lactobrevin
Lacxtobacillus acidophilus	Acidolin
	Acidophillin
	Lactobacillin
	Lactocidin
Lactobacillus plantarum	Lactolin
Lactobacillus bulgaricus	Bulgarican
Bifidobacterium bifidum (some strains)	Bifidin

The fact that bifidobacteria superbly control and deactivate other (often disease causing) bacteria, has led many experts to infer the presence of such antibiotic substances relating to the active presence of these most important friendly bacteria.

Drs Shahani and Chandan were the researchers who first isolated the antibiotics Bulgarican and Acidophilin, and they are on record as saying that these are, 'exceedingly active against a wide variety of . . . organisms which included pathogens and non-pathogens'.

> **Note:** Dr Shahani also makes it clear that the production of antibiotics by strains such as *L. acidophilus* DDS-1 and *L. bulgaricus* only takes place when these have been cultured in a milk medium.

Among the pathogenic (disease causing) bacteria which acidophilin inhibits in the human body are: *Salmonella typhosa* and *schottmuelleri; Shigella dysenteriae* and *paradysenteriae; Pseudomonas fluorescens* and *aeruginosa*; *Staphylococcus aureus*; *Klebsiella pneumoniae* and *Vibrio comma*.

> **Note:** see page 99 for details of a potentially lethal connection between *Staphylococcus aureus* and *Candida albicans*. Please also remember *Klebsiella*: we will discover some dramatic facts about its unwelcome presence in the GI tract in the next chapter.

Not all of the control factors which friendly bacterial strains exert over harmful bacteria (and non-harmful yet rival bacteria) relate to their anti-microbial production. For example, in the case of *Lactobacillus acidophilus*, Drs Shahani, Fernandes and Amer inform us: 'Its anti-microbial activity is due to production of lactic acid, hydrogen peroxide and antibiotics.' It seems that it is the combination of these control elements which produce the benefits of which we can take advantage.

Acidolin, one of the natural antibiotics, has been investigated by the US Department of Food Science and Nutrition, in Columbus, Ohio, where researchers E. Mikolajcik and I. Hamdan carefully extracted the substance by painstakingly repeated chemical and filtration processes. They found in the end, 'a yellow viscous liquid' which was stable to heat and long-term storage, very acidic and active against a wide range of organisms, but not against other lactic acid bacteria. It was found to be non-toxic to human tissue culture cells.

Researchers into natural antibiotics produced by the lactobacilli report that different strains vary widely in their production of these. For example, they found the powerful antibiotic acidophilin was produced by the strain of *Lactobacillus acidophilus* known as DDS 1, and bulgarican by *Lactobacillus bulgaricus* DDS 14.

Of major importance to the antibiotic producing bacteria are variables such as the acidity and temperature in which they live, as well as the food on which they are fed (milk was always best). These factors strongly affect their ability to produce antibiotics: indeed, when cultured on mediums other than milk these same strains may be able to produce no antibiotics at all.

Drs Shahani, Fernandes and Amer issue several guidelines and precautions which anyone aiming to use such cultures for health benefits should take note of:

1. They point out that benefits found with one strain may not apply to other strains of the same organism.
2. They warn that commercial preparation of cultures which are shown to be effective in laboratory conditions may not contain sufficient numbers of viable and potentially active organisms to be of any therapeutic or preventive benefit.
3. They warn that these tiny creatures are 'fastidious' in the requirements they have in order to grow and produce their useful antibiotics and other by-products, and that their dietary patterns (and, by implication, yours) can influence this markedly, as can the presence of alcohol and, quite obviously, medical antibiotics.

Among the more obvious general benefits which we derive from such natural antibiotic production are the increase in shelf-life due to the resulting inhibition of spoilage bacteria, as well as the potential for direct destruction or inhibition of disease-causing micro-organisms, or the stopping of their toxin production inside ourselves.

Testing the antibiotic properties of these bacteria

In laboratory studies, a given quantity of an antibiotic such as acidophilin is placed in a culture dish in which has been growing one or other of the disease causing micro-organisms known to be sensitive to it. After a set period of time the area around the antibiotic, which is by this time clear of pathogenic bacteria (they have been killed by the antibiotic) is measured. This is called the 'zone of inhibition' and clearly the larger the dimensions of this zone, the greater we can assume the antibiotic potential to be against the particular organism being tested. When acidophilin and bulgarican were tested against some of the most unpleasant and dangerous food-borne pathogens the following results were obtained. (Remember that the greater the zone of inhibition, given in millimetres in the table below the more powerfully did the natural antibiotic destroy the disease-causing bacteria):

Pathogenic Bacteria	**Acidophilin**	**Bulgarican**
Bacillus subtilis	45mm	44mm
Clostridium botulinum	37mm	38mm
Clostridium perfringens	31mm	33mm
Eschirichia coli	44mm	40mm
Proteus mirabilis	39mm	45mm
Salmonella enteridis	42mm	39mm
Salmonella typhimurium	44mm	39mm
Shigella dysenteriae	30mm	not tested
Shigella paradysenteriae	30mm	not tested
Staphylococcus aureus	35mm	38mm
Staphylococcus faecalis	31mm	39mm

(adapted from Fernandes, Shahani and Amer)

When bifidobacteria were tested against these same pathogens the results were not as dramatic, but were nevertheless impressive. The bifidobacteria have a powerful anti-bacterial potential and seem to exert their influence by means of the production of organic acids, such as lactic, acetic and formic. Different strains produce different acids in varying quantities. Acetic acid has the strongest effect on potentially pathogenic bacteria, such as shigella and salmonella. The effectiveness of these methods of control will become clear when we examine their role in enteric infections. A variety of different strains of bifidobacteria have been studied for such properties, with strain 1452 emerging as the clear winner in terms of its overall ability to keep pathogens down.

Interestingly, when this strain (1452) was cultured on varying foods (either skimmed milk or a broth) it showed wide fluctuations in its abilities, with that cultivated on milk being superior. These differences, in which those grown on skimmed milk were often seen to be twice as powerful as those cultivated on broth, are indicated in the table below. This again emphasizes the importance of ensuring a high potency, active culture, grown on a suitable medium (i.e. milk) if you intend to supplement with a friendly bacteria. It also emphasizes the importance of your own diet being suitable for the health and growth of the culture once it is living inside you. We suggest that you re-read the notes in Chapters 1 and 2 which discuss bifidogenic factors in the diet that can positively enhance the vigour of these little friends.

Pathogenic bacteria	Zone of inhibition of *B. Bifidum 1452*	
	Skimmed milk	*Broth*
Escherichia coli	20mm	16mm
Bacillus cereus	22mm	16mm
Salmonella typhosa	12mm	8mm
Shigella dysenteriae	11mm	not tested
Micrococcus flavus	25mm	18mm
Staphylococcus aureus	23mm	14mm
Pseudomonas fluorescens	18mm	11mm

(Adapted from Anand, Srinivasan and Rao)

The researchers into these anti-bacterial properties of certain strains of bifidobacteria tell us that these powerful inhibiting effects on pathogenic bacteria were only observed in an acid medium below pH5.5. Above that level, in a less acid medium, the control factors, whatever they are, produced by bifidobacteria, are not able to inhibit these pathogens.

Examples of antibiotic and anti-bacterial use of friendly bacteria

1. The Polish study referred to in the previous chapter involved

16 children with salmonella dysentery, and 15 with shigella dysentery, who were treated with acidophilus milk. After the initial treatment, 7 of those with salmonella and 10 with shigella were clear of symptoms, and with long-term use of the acidophilus milk all cases were cleared of infection and symptoms.

2. The preventive capabilities which the antibiotic content of such cultures offers is made evident in a six-month long Japanese study of soldiers. A thousand men were divided into two groups, one of which received a fermented product called yakult (using the micro-organism *L. casei*, formerly known as the *Lactobacillus acidophilus* strain shirota).

After six months 50 of the 500 not receiving the cultured product were found to be carriers of the pathogenic micro-organisms salmonella or shigella, and 55 of the 500 had dysentery, whereas none of the 500 soldiers receiving the yakult had any symptoms or were carriers.

3. In animal studies it was found that rats which had a high yogurt content in their diets were resistant to *Salmonella enteridis* and survived such infection better than did those rats not eating yogurt.

4. The antibiotic potentials of acidophilus were studied in Sweden at the world-famous Karolinska Institute. The researcher, Dr Alm, was interested in helping people who had been infected by salmonella to clear their systems of this common cause of food poisoning and 'traveller's tummy'. The purpose of the study was to see whether the taking of *Lactobacillus acidophilus* would shorten the period during which the person (child or adult) remained a carrier. Usual treatment is medical antibiotics, but there is no guarantee that this will actually eliminate the carrier state.

Treatment involved the drinking (ideally at one time) of not less than 500 millilitres of an acidophilus milk which contained not less than 6 billion micro-organisms per millilitre.

Compared to a group of control patients who consumed no cultured milk products, those 'carriers' who started on the acidophilus programme soon after their diagnosis (by stool culture) had a much shorter period of infection, with none being carriers after six weeks.

5. Research at North Carolina State University, by Drs Gilliland and Speck, examined different strains of acidophilus grown in different mediums and under different conditions, and used at

different levels of acidity in order to assess the effects of these many variables on their antibiotic potentials.

The complexities of the study need not concern us; the results, however, should. They showed that *Lactobacillus acidophilus* was antagonistic to the growth of *Staphyloccus aureus, Salmonella typhimurium* and *Clostridium perfringens* (the pathogens tested) to a very great degree. In some instances there was up to a 98.2 per cent degree of inhibition, which means the pathogen was virtually wiped out.

Such benefits are most notable when *L. acidophilus* comes into contact with the pathogen early on in its growth stages.

Researchers believe that the control of pathogens results from a combination of the acidity created by acidophilus, its production of hydrogen peroxide and, of course, its antibiotic secretions.

In order to see whether acidophilus could exert these effects in anaerobic conditions, such as would be found in the colon, specialized experiments were undertaken and these did indeed prove that the antibiotic effects were still strongly evident even when oxygen was absent.

6. Strains of bifidobacteria (notably fragilis) were studied in Japan in relation to their abilities to protect against shigella bacteria infections of the intestine.

Before examining this important research effort we should familiarize ourselves with the variations in the way different micro-organisms create intestinal havoc when they infect us.

Disease	**Infecting organism**	**How and where damage is done**
Cholera	*Vibrio cholera*	Infecting bacteria attach to mucous lining of small intestine, multiply and produce the toxin which produces symptoms.
E. Coli enteritis	*E. Coli* strain	
Bacillary dysentery	shigella	Bacteria attach to and penetrate lining of large intestine, multiply and damage local tissues, and sometimes produce toxins.
E. Coli enteritis	*E. Coli* strain	
Campylobacter enteritis	*Campylobacter jejuni*	As above in small or large intestines.
Salmonella enteritis	salmonella	Bacteria attach to mucosa, of small (or, in case of *S. typhi*, large intestine which they penetrate to

Typhoid fever	*Salmonella typhi*	reach underlying tissues, where they multiply and may spread systematically.
E. Coli enteritis	Enteropathogenic enteritis	Bacteria damage minute projections on intestine wall (microvilli) and damage cells of the region but do not penetrate the mucous lining. They induce disease by impairing digestion and absorption of nutrients, and may also produce toxins, both in small and large intestines.

In his research, Dr Ninkaya (from whose work the table above is adapted) studied the inhibitory effects of bifidobacteria on shigella bacteria derived from patients with bacillary dysentery.

These highly active disease bacteria were placed in dishes in which were growing, in a culture medium, human intestinal epithelial (lining) cells. By assessing how much damage the shigella bacteria could produce on these defenceless cells, on their own and accompanied by various concentrations of bifidobacteria, it becomes possible to specify the degree of protection which could be offered by the friendly bacteria.

In four different experiments it was seen that:

1. When shigella was added to the cell dish with no protection offered, approximately 12 per cent of these cells became infected.
2. When shigella and *Bifidobacterium infantis* were added simultaneously only 2 per cent of the cells became infected.
3. When shigella was allowed to infect the cells 2 hours after *Bifidobacterium infantis* had been added only a 1.6 per cent infection rate was observed.
4. When the cells were protected for 2 hours by *B. infantis* and were then washed and subsequently infected with shigella, 4.4 per cent of the cells became infected.

These experiments clearly show that bifidobacteria interfere with invasion, and probably multiplication, of shigella organisms. Notice that the most protection was achieved when the bifidobacteria were present *before exposure to the disease causing organism*, a clear indication of the value of always having a healthy GI tract, populated with just such a defensive army of friendly bacteria.

Other experiments involved the mixing of different concentrations of the protective bacteria, and this clearly demonstrated that the higher the concentration of bifidobacteria, the greater the protection offered.

When other strains of bacteria, such as *Bifidobacterium pseudolongum* and *breve*, *Streptococcus faecalis*, and various benign strains of *E. coli* were tested to see to what degree they could inhibit shigella activity (in culture dishes with live cells) all showed an ability to inhibit the degree of infection, but none as effectively as *Bifidobacterium infantis*.

Is the protective factor something the bifidobacteria secretes?

In order to find out whether the protective substance against shigella was in the *B. infantis* bacteria, or related to something it secreted or synthesized, a subsequent study was carried out involving adding fluid in which the *B. infantis* bacteria had previously been growing to the dishes in which the human intestinal cells were being cultured. The infantis bacteria had been separated out from this fluid by centrifuging them (rapid spinning) or filtration, leaving behind what is called culture supernatant, which no longer contains any actual infantis cells. This fluid (the supernatant) had a strong inhibitory effect on shigella's ability to infect cell cultures, proving, in the words of the researcher, Dr Ninkaya, 'the presence of some active factor(s) in the culture supernatant that protects cells from invasion by and proliferation of *Shigella flexneri* organisms'. The higher the concentration of supernatant, the stronger the protection.

These and similar studies indicate strongly that the bifidobacteria, or at least some strains of these large intestine inhabitants, offer non-specific protection against infection by pathogenic micro-organisms.

Reference to the table above will remind you of the way shigella infects in bacillary dysentery, by penetration of the bowel mucosa and subsequent multiplication and toxin production. If, as these studies indicate, the actual infection of mucous membrane cells can be stopped by the as yet mysterious secretions of *Bifidobacterium infantis*, none of the subsequent damage would take place.

Just what these (possibly antibiotic) protective substances are remains unresolved, and at the time of publication of this important Japanese research (1984) it was stated that further investigations were under way to try to identify just how this protection is produced. In the meantime, it is clear that the health of the bifidobacteria inhabitants of the GI tract is deserving of some consideration, and that supplementation with bifidobacteria (ideally in a viable powder form with guaranteed potency linked to a stated expiry date) and bifidogenic factors (see Chapters 1, 2 and 14) in the diet, are useful strategies which can add to the protection of this vital region.

As we will see in later chapters, the way commercial preparation of the friendly bacteria takes place may involve the rapid spinning (centrifuging) of the culture to separate the bacteria from their growth medium (supernatant). Japanese experts such as Dr Ninkaya (as well as European authorities such as Dr Rašić) clearly state that this separation method weakens the ability of the friendly organisms to inhibit disease-causing bacteria, as well as weakening the viability of the friendly bacteria by damaging their cell-walls, as we shall see. Filtration is therefore to be preferred even though it still strips the supernatant away.

It seems that the supernatant (the material in which the friendly bacteria are grown) has itself specific inhibitory powers over pathogens such as shigella. It pays, therefore, to seek products which are not only of high potency (guaranteed, of course, with a clear indication of how long that potency will last) but which have not been centrifuged or ultrafiltered during their production for supplementation.

Some manufacturers of acidophilus products (usually those manufacturers who employ centrifuging methods) argue that the bacteria can produce a new supernatant after they reach the digestive tract. Thus far there have been no tests or studies which support this line of thought. We believe that it is vitally important, when you buy these microbial products, that you ensure that you are obtaining not only viable high potency cells, but that these are undamaged and contain their supernatant as well.

Further Japanese research on other pathogenic organisms

When people are severely ill, as in cases of leukaemia, and are treated with chemotherapy, drastic changes occur in their GI tract. The intestinal flora which is usually sparse in pathogens such as klebsiella, citrobacter and *Proteus vulgaris* becomes densely populated with these aerobic bacteria as well as anaerobic cultures such as bacteroides which increase significantly in number. A proliferation of *Candida albicans* is usually also observed.

This sort of overgrowth provides an ideal situation for testing the efficacy of the bifidobacteria and lactobacilli in their role as controllers or destroyers of pathogenic micro-organisms.

Before starting the study researchers did the following:

1. Isolated, identified and estimated the numbers of the intestinal bacteria of 56 leukaemia patients.
2. Identified and classified the level of candida overgrowth.
3. Tested for toxins created by pathogenic bacteria, in the blood and urine.

Then, instead of the water usually consumed by these patients, they substituted milk which contained high potency levels of bifidobacteria and *Lactobacillus acidophilus*. There were not less than 10 million organisms of each of these per millilitre of milk, and 200 millilitres were consumed daily by each patient for three months (a total consumption of not less than 2 billion of each organism daily).

Amazingly, the intestinal balance, in terms of relative numbers of potential bacterial pathogens, returned to normal on this regime.

Candida albicans levels were very high in these patients (we will deal with candida in detail in the next chapter: it is mentioned here as part of this particular study). There were 100 million candida organisms per gram in four patients, 10 million organisms in six patients and 1 million organisms in 4 patients, who subsequently received the bifidobacteria/*L. acidophilus* treatment.

Three patients at each of these three levels of infection (100, 10 and 1 million organisms) were kept as controls, that is they received no bifidobacteria/*L. acidophilus* treatment.

There were obviously no changes at all in the group (controls) who had received no friendly bacteria treatment, but significant improvements amongst those who had, there no longer being any patients with concentrations above 1,000,000 micro-organisms per gram of faeces.

S. faecium and enteritis

European research into the treatment of enteritis shows that a particular strain of *S. faecium** (S.F.68) has useful potentials in treating this condition in both children and adults, as well as having some ammonia detoxifying qualities. However, most of the studies referred to, conducted in Switzerland and Italy, were of short duration (6 to 10 days) and were always under clinical conditions (not self-medication) during acute episodes of infection.

We retain our reservations as to the long-term use of *S. faecium* by the public, and support the doubts expressed in this regard by Professor Rašić when he states: 'There is scarce scientific evidence as to the possible beneficial effects of these bacteria in the human intestines; especially when compared with *L. acidophilus* and bifidobacteria, or even with transient yogurt bacteria.' Once *S. faecium* drives out pathogenic bacteria and gains a hold on binding sites it could prove very difficult to dislodge. Drs Susan Minney and Nigel Plummer, in personal correspondence with the authors state: '9 per cent of all hospital patients have secondary infections where *S. faecalis* is implicated . . . Due to the inherent pathogenicity of the streptococci, their use as dietary supplements should be discouraged.' If this strain of *S. faecium* is used therapeutically we are of the opinion that it should be for short periods only, under clinical conditions (hospital etc.) or at least under supervision from a qualified health professional, and not in the sort of 'cocktail' probiotic combination currently being marketed in Europe and the US.

According to Bergy's Manual of Systematic Bacteriology, Volume 2; . . . a new genus *Enterococcus* has recently been proposed to include the species described in this section (*Streptococcus faecalis*, *S. faecium*, "*S. avium*" and *S. gallinarum*).' (For this reason you will find *S. faecium*, *S. faecalis* etc. interchanged with *E. faecium*, *E. faecalis*, etc.).

Lactobacillus control of pathogens

The food industry is naturally anxious about bacterial

contamination of its products. American research has focused on the benefits which the antibiotic potential of *Lactobacillus bulgaricus* has to offer.

Commercially prepared foods such as refrigerated tuna fish, potato salads, refrigerated minced (ground) beef (used in the hamburger industry) are frequently (a notable research study says 'usually') contaminated with large numbers of bacteria. Even at refrigeration temperatures, staphylococcus, pseudomonas and achromobacter cause spoilage of such foods.

A 1983 Florida study by Drs Abdel-Bar and Harris looked at the potential for preventing such contamination using lactic starter cultures such as *Lactobacillus bulgaricus*. The effect of three different concentrations of this culture on the type of foods described was observed for five days with the food at 43°F/6°C.

There was total growth prevention of the spoilage organisms being tested for when high concentrations of *Lactobacillus bulgaricus* were used (between 1.4 million and 5.7 million organisms per millilitre of concentrate).

The researchers believe that a combination of the production of acid or hydrogen peroxide (H_2O_2) is the means whereby *L. bulgaricus* achieves its protective control of pathogenic organisms. These methods are suggested as being potentially useful in the food industry and where meat such as minced (ground) beef is being transported or stored. We discuss the enormous problem of food poisoning by such organisms as salmonella in greater detail in Chapter 12.

The evidence discussed above shows that if enough of the friendly bacteria are supplemented, in food or other forms, in a viable and active state, they can offer our GI tracts, and therefore us, enormous protection from infection as well as aiding in ridding the body of pathogenic organisms. In a later chapter we will examine some of the evidence relating to the control of bacterial changes in the vagina and the relationship this has with the friendly bacteria which live in that region.

Chapter 6

Osteoporosis and ankylosing spondylitis: two unexpected consequences of dysbiosis

In the previous chapters, as we have become familiar with the sort of nutritional and environmental factors which can alter the balance of the sensitive ecosystems operating inside our GI tract, we have gained an awareness of some of the consequences of changes, such as acute diarrhoea. The treatment best suited to such episodes is reasonably obvious and should clearly include repopulation of the flora of the gut as rapidly as is possible.

Some of the ways in which this has been done in clinical studies, using acidophilus milk, and supplementation of various combinations of the friendly bacteria, has been outlined. In Chapter 17 we will present suggestions as to ways in which such nutritional supplementation might be used by you in similar conditions as part of an attempt to bolster the natural defence systems of the body. We will also give guidelines for situations where the involvement related more to vaginal or genito-urinary problems, or where infection results from viral or yeast activity, rather than changes in bacterial populations.

Osteoporosis

Some consequences of altered bowel flora are far less obvious, though. Recall that when the normal flora is disrupted, by antibiotic therapy for example (see Chapter 4), changes occur in which excessive amounts of the female sex hormone oestrogen are eliminated instead of being recycled. This results in low levels of this important hormone being present in the bloodstream.

Osteoporosis, or 'brittle bone disease', is now rampant in Western industrialized countries, affecting millions of women and causing untold suffering and disability, as bones such as those of the upper thigh snap like dry twigs, and spinal fractures occur without warning. This veritable epidemic relates to a number of causative factors, and one of these is the change in bone metabolism when oestrogen levels are lowered (such as occurs naturally with the menopause).

Women are constantly being advised to take calcium supplements in the years leading up to, and beyond, the menopause, in order to forestall the effects of calcium loss, which is thought to be largely the result of oestrogen's natural decline. *If, however, oestrogen's presence is low anyway, due to altered bowel flora activity, regardless of the natural decline due to age, bone loss problems are bound to be exacerbated and to start often long before the menopause begins*.

Declining oestrogen production after the years of fertility tends to disturb the calcium balance, leading to bone porosity and brittleness. Instead of calcium being taken up by bone at this time, maintained by oestrogen and parathyroid hormone levels, it is being 'leached out' of the bones and deposited in the soft tissues (it may even result in kidney stones).

The taking of calcium and magnesium supplements (these two minerals are intimately related and should usually be supplemented together in a ratio of 2:1) is of proven value in preventing the worst aspects of this decline, as is regular weight bearing (aerobic) exercise.

We suggest that for many women a strategy should be adopted in the years *before* the onset of the menopause involving replenishment of the bowel flora, thus enhancing oestrogen levels which might be in decline due to dysbiosis of the GI tract. Such an approach could be of great importance, as it would slow the decline in this vital hormone's influence in the critical middle years.

Another contributory dietary factor to osteoporosis, apart from inadequate intake of calcium and magnesium, often involves a high meat content in the diet. This has been shown (M. Werbach, *Nutritional Influences on Illness*, Thorsons 1989) to severely compromise calcium levels in the body (remember that a high protein intake — especially when accompanied by meat fat — also disturbs the functional behaviour of the intestinal flora). Women who consume excessive quantities of meat usually have demonstrably greater degrees of osteoporosis.

It seems that there may well be a connection between

the changes in the flora, which such a diet produces (see Chapter 3), and the negative calcium status which occurs when meat in particular, and excessive protein in general, are a feature of the diet.

Low oestrogen levels which develop after antibiotic treatment, or through general dietary abuse (high fat levels for example) are certainly a factor in osteoporosis. Anyone with a tendency to this sort of bone loss would be well advised to ensure a healthy intestinal flora as part of any plan to ensure healthy bones. The classic characteristic of women who are most prone to osteoporosis include being slim, white, on a high protein diet, a smoker, having low levels of exercise and approaching, or having passed, the menopause.

In addition to calcium/magnesium supplementation, adequate exercise and a balanced high fibre diet with enough, but not excessive levels of, protein, this would call for regular consumption of cultured milk products (real cultures, that is, not commercially acidified milk!) and/or inclusion into the diet of concentrated high potency lactobacilli and bifidobacteria cultures in powder form (with guaranteed potency and expiry date) so that the health of the bowel is ensured.

'Bamboo spine'

Another consequence of bowel dysbiosis, (i.e. altered ecology of the flora of the GI tract, incursion of undesirable bacteria, and all the obvious consequences of such changes) includes one most surprising and unwelcome possibility.

People who develop ankylosing spondylitis (AS) suffer with an ever-increasing level of spinal stiffness and pain. Ultimately, the spine becomes fused into a bent position, often to such an extent that the individual cannot look ahead without extreme difficulty. The damage which occurs in the spine relates to a gradual encroachment of each spinal segment or vertebra to the ones above and below it. Bridges of bone gradually turn the previously flexible spine into a rigidly fixed unit not unlike a section of a bamboo pole; only not straight, but bent.

The bones of the pelvis may also fuse, resulting in pain and great difficulty in sitting comfortably. Other joints may in time become affected by the same process, and there is often a major degree of involvement of lung and heart dysfunction, partly due to the mechanical crowding which the progressive condition causes.

In the early stages of the disease, symptoms such as depression, chronic fatigue and severe weight loss are common, along with debilitating pain.

There is no medical treatment for this disease, apart from the symptomatic use of anti-inflammatory drugs and painkillers, neither of which have any effect at all on the progressive element of this intermittently active condition, in which periods of inflammation are interspersed with quiescent times. The disease severely affects over 80,000 people in the United Kingdom and an estimated 750,000 people less acutely. The figures for the USA indicate that hundreds of thousands of people are badly affected, with millions of others having related symptoms which they may regard as 'back pain' or 'a bit of arthritis'.

Recently, however, some fundamental research has revealed a quite startling bacterial connection with this crippling affliction. AS is known in medicine as an 'auto-immune' disease. This means that the symptoms result from what appears to be an attack on aspects of the body by its own immune (defence) system.

Other conditions which have been so identified include ulcerative colitis and rheumatoid arthritis, and there seems to be a very real link between these destructive conditions and alterations in the popularions of the GI tract, involving excessively high populations of some of the less friendly bacteria whom we have met in previous chapters, such as klebsiella, shigella and versinia.

Bacterial overgrowth: The link between rheumatoid arthritis and AS?

It has been discovered that some human tissues have remarkable similarities to the biochemical constituents of certain bacteria. Human tissue can be assigned to different 'types' depending upon its peculiar biochemical and cellular characteristics. One such tissue-type has been given the designation B27, and this is the type which is found in fully 98 per cent of people suffering from ankylosing spondylitis.

Dr Alan Ebringer, of King's College Hospital, London, wondered whether the body's immune system might not be making a mistake in its efforts to get rid of an undesirable bacteria by attacking its own tissues instead, initiating self-destructive processes such as occur in auto-immune diseases. His research team carefully began the process of typing different bacteria retrieved from the bowels of AS patients, trying to find one which closely matched the B27 tissue-type characteristics. They found that the bowel bacterium which most closely matched this tissue was klebsiella. The immune system could, they believed, easily mistake its own tissues for this and begin to attack itself in an attempt to remove the organism. Once this link was found, tests began in which the relative numbers of klebsiella in the faeces of people with AS were examined during different phases of the condition. It was discovered that in the active stages, when inflammation was at its peak, klebsiella levels in the stools, and antibody levels to klebsiella in the blood, were very much higher than in periods when the condition was relatively calm.

Tests also revealed that men have far higher levels of klebsiella than women, and this corresponds to the fact that three times as many develop AS as do women.

Further research is under way to identify other bowel bacteria which may possess similar characteristics to other tissue-types, found in people with different auto-immune diseases (such as myasthenia gravis, lupus erythematosus, motor neurone disease, pernicious anaemia and forms of diabetes). An initial breakthrough has been achieved with rheumatoid arthritis, another auto-immune disease. Teams of researchers at hospitals in the UK established that the tissue-type most frequently found in such patients was HLA-DR4, and that the bacteria which is most similar to this is proteus, a common cause of urinary tract infections.

Women have a far higher incidence of urinary tract infection than men, and they also suffer rheumatoid arthritis twice as frequently as do men. When the levels of antibodies to proteus in the blood of people with rheumatoid arthritis was compared with that found in healthy people, it was discovered to be far higher. This observation has now been confirmed in Ireland, Finland and other research centres, where the possibility of a link between bacterial overgrowth and auto-immune disease is attracting great interest.

The King's College dietary treatment plan for AS

Having established the link between klebsiella and AS the doctors at King's introduced a dietary strategy with AS patients as well as usual normal symptomatic medication. This is basically a 'low-starch diet', and Dr Ebringer has explained it thus:

'Klebsiella thrives on a diet rich in starch. If you cut out starchy carbohydrates such as rice, potatoes and flour products, then you reduce the number of klebsiella in the gut, and subsequently the production of antibodies to the bacteria which cause the inflammation.'

Patients are given instructions to cut out bread, pasta, cereals of all sorts, rice and potatoes, as well as sugary foods. They are unrestricted in eating vegetables, fruit, eggs, cheese, fish and meat.

Does it work? Over 200 patients have so far been through this programme and Dr Ebringer claims that the majority have had their disease process halted. But it is not a 'quick' cure by any means. One patient is quoted as saying: 'Once I stuck to the diet religiously I noted a real improvement after six months or so. Movement became easier and the lethargy and depression lifted. The best way I can describe it is that after years of pain and stiffness I suddenly feel "well-oiled".'

Do we agree with these concepts

We feel that the King's College team have seen part of the truth, and have made a major contribution to our knowledge. However, they have not gone far enough. We have seen in earlier chapters just how the bowel ecology can be helped by repopulation by the major friendly bacteria such as *L. acidophilus* and the bifidobacteria, as well as by use of transient visitors such as *L. bulgaricus*. If klebsiella is present in over-abundance, this repopulation is surely the first strategy to adopt (along with a bifidogenic diet).

If klebsiella is active in the body, in regions beyond the closed world of the GI tract, it must have found a means of entry to the bloodstream, and this is almost certainly related to the damage which *Candida albicans* is capable of doing to the mucous lining of the gut, as we will see in the next chapter.

An anti-candida aproach is therefore called for, and we will see, one of the best ways of keeping candida down is a low starch diet. Are the King's doctors in fact unknowingly achieving their results by partially controlling candida, as well as klebsiella? We would suggest this to be the case, and would urge the use of friendly bacteria to further assist in the control of the probable bacterial trigger to AS, by helping to re-establish a normal bowel flora, something which would keep both klebsiella and candida in their place.

Dysbiosis, the alteration for the worse of the ecology of the digestive tract, is here shown to be at the very forefront of causes of major degenerative and destructive diseases. Probiotics, the restoration by sound supplementation and diet of this ecosystem offers a major contribution to both recovery and, of course, to prevention.

In the next chapter we will reveal further mysteries in this puzzle, as we investigate the ways in which the friendly bacteria can help to control rampantly destructive yeasts and hostile viruses.

Chapter 7

Antibiotic action in the urinary tract

Vaginitis

Very nearly 100 years ago Albert Doderlein, a famous German obstetrician, described the characteristics of what came to be called the Doderlein bacillus, a normal and protective inhabitant of the vagina. When he described this in 1892, its clinical importance was not understood and at first it was even thought that the bacillus was capable of causing disease. Subsequently the Doderlein bacillus was found to be our now familiar friendly bacteria, *Lactobacillus acidophilus*.

As research continued, it became clear that not only was it not harmful, but its abundant presence was indicative of good health in the region. Doderlein found that the secretions of the vagina could be roughly divided into three groups, in terms of their significance as health indicators.

Grade one includes about 40 per cent of non-pregnant women. The secretions are acid in nature and have almost exclusively the Doderlein bacillus in residence.

Grade two have a mixture of Doderlein and other bacteria (about 18 per cent of women examined).

Grade three have hardly any Doderlein bacillus (*L. acidophilus*) and a vast number of pathogenic micro-organisms such as diptheroids, streptococci, micrococci etc. and in these women (42 per cent of those tested) the secretions were found to be alkaline.

When healthy young women (i.e. those with no obvious gynaecological disease) are tested 100 per cent are found to have resident cultures of Doderlein bacillus in the vagina.

One approach to problems such as vaginitis has been to culture healthy Doderlein bacilli (*L. acidophilus*) taken from women free of disease and to treat such conditions by implanting the flourishing colonies of these friendly bacilli into the vagina, allowing them to re-establish dominance in the region.

A Spanish study involved 49 women with vaginitis in which the only treatment offered was a daily application of pure cultures of Doderlein bacillus. There was immediate relief of symptoms, and evidence of the replacement of staphylococcus, diplococcus and streptococcus (and other pathogenic bacteria) with Doderlein bacillus. In all cases the acidity shifted from between pH5 and pH6 to a more acid pH4.

A summary of 58 cases of various forms of vaginitis treated with pure cultures of Doderlein bacillus (*L. acidophilus*) showed:

Condition	Number	Cured	Symptoms relieved	Failure	Recurred
Non-specific vaginitis	19	95%	5%	0	0
Monilia vaginitis	25	88%	12%	0	0
Trichomonas vaginitis	8	87%	0	13%	25%
Trichomonas and monilia vaginitis	6	100%	0	0	0

(adapted from Butler and Beakley)

The spectacular success achieved by this simple approach, in so many cases of this distressing, painful and socially awkward condition, leads to the obvious question: why is it not universally used? There is no satisfactory answer, although the fact that treatment requires an initial cultivation of healthy Doderlein bacteria, a somewhat laborious task, and the total lack of any commercial potential in this method of treatment for ethical drug houses, may have much to do with the answer.

It is worth remembering that one common factor in recent years, leading to an alteration of the acid/alkaline balance in the vagina, relates directly to the use of the contraceptive pill. Anyone regularly using this form of contraception should consider using nutritional tactics, involving repopulation of acidifying lactobacilli, as part of an attempt to minimize the distruptive effects of 'the pill'.

Further evidence

The successful use of Doderlein (*L. acidophilus*) cultures in normalizing the acidity of the vaginal region and allowing repopulation by these normal residents has been shown in many studies, as has the application of *L. bulgaricus* in such problems.

- In laboratory experiments Dr Shahani has shown that acidophilin (extracted from *L. acidophilus*) produced a 50 per cent inhibition of 27 different types of bacteria, 11 of these were common disease causing agents.
- In another research study the presence or absence of lactobacilli was established in 12 patients with vaginitis (associated in this instance with gardnerella) both when the patients had symptoms (itching, burning etc.) and when they were symptom-free. The gardenerella organism was easily found both when symptoms were strong and when there were no symptoms.

  However, lactobacilli were more commonly found when the women were symptom-free, but were usually absent when symptoms appeared. This suggests that the overall disturbance of the ecology of the region allowed the more or less permanent presence of the disease-causing organism gardnerella. When lactobacilli could repopulate the region the symptoms were kept in check; when for one reason or another (remember, lactobacilli are extremely sensitive to stress) the lactobacilli were depressed in numbers, or absent, the symptoms reappeared.

  This sort of research has been repeated many times, and has led some doctors, such as Dr C. Spiegel, to conclude that when there are few (or no) lactobacilli present in the vaginal region in cases of irritation and inflammation of the vagina, an assumption can be made that bacterial vaginitis exists, whether or not any pathogenic bacteria can be isolated.
- German research involved in the application of doses of 7 billion lactobacilli per half millilitre into the vaginas of women with non-specific vaginitis. Of 94 patients treated in this way, 80 per cent were cured or returned to a symptom-free state, with evidence of a decrease in the

pathogens, the growth of normal lactobacilli and a return to normal acid levels. Since this method offers a strong chance of prevention of further infection, unlike conventional drug therapy, it would seem to offer a decided advantage.

- When 444 women with *Trichomonas vaginalis* were treated with *Lactobacillus acidophilus* (brand name Solco-Trichovac/Gynatren) of the 96 per cent who were rechecked a year later fully 92.5 per cent were clinically cured. The remainder (7.5 per cent) showed a positive culture of acidophilus.

  This clearly demonstrates the ecologically sound concept of repopulating the region with bacteria which can keep the conditions required for health, and that this method has a lasting benefit, not just a short term one.

Paul Reilly ND, writing in *The Textbook of Natural Medicine* (JBCNM, Seattle, 1987) discusses the use of *Lactobacillus acidophilus* in treating vaginitis. He says: 'Whenever there is a disturbance of the normal vaginal flora, re-establishment of these organisms is important. This may be accomplished by the insertion of live lactobacillus culture yogurt (careful reading of the labels is important since most commercially available yogurts do not use lactobacilli) but a more efficient and less messy method is to insert one capsule of lactobacillus into the vagina twice daily for one or two weeks.'

The advice regarding insertion of yogurt is based on traditional folk medicine. That regarding insertion of acidophilus capsules presupposes a high potency, active acidophilus culture in the capsule.

A little (quarter teaspoonful) of one of the active high potency acidophilus cultures, now readily available in powder form, mixed in a little live yogurt and inserted into the vagina and kept in place with a tampon or sanitary napkin, is one method which quickly helps in repopulation on a local level, and is very soothing if the area is itchy or burning.

We suggest, though, that empty capsules are obtained from a pharmacy and that these be filled with a high potency acidophilus powder (guaranteed until a specific expiry date), prior to insertion into the vagina. Four 500 milligram capsules or 2 one gram capsules should be inserted every night for ten days to a fortnight. This will usually deal with local infection and inflammation, and should of course be accompanied by the taking by mouth of not less than a half teaspoon of high potency acidophilus powder, three times daily in tepid water, away from mealtimes (see last chapter). As we will see when we look at the problems associated with *Candida albicans*, in Chapter 8, vaginitis, when it is caused by this yeast, can also be dealt with by variations on the theme of lactobacillus insertion.

> Note: We believe, after diligent research, that capsules of (for example) *L. acidophilus*, which are filled at the time of manufacture, lose their potency rapidly, especially if they are not constantly refrigerated. Independent testing has shown that there is often little viable activity in bacterial cultures thus prepared, however high their potency before being packed. Packing your own high potency, powdered (refrigerated) cultures into capsules, for vaginal insertion, is a slightly irksome procedure, but it really works.

Until a manufacturer produces an encapsulated acidophilus product which carries a guaranteed potency with an expiry date, you, the consumer, may lose confidence in the concepts and methods we have presented to you, as a result of frequently unsatisfactory results.

Urinary tract infections in women

In many instances, urinary tract infections relate to *Candida albicans* overgrowth, and this will be dealt with in Chapter 8. In this chapter it is the possibility of bacterial involvement which concerns us. The distress caused by an over-frequent need to urinate and the discomfort so often associated with this problem is such that the afflicted woman may feel a social outcast.

Conventional medical treatment involves the use of antibiotics, which may give short term relief, but often does not. Recurrence commonly occurs as soon as antibiotics are stopped, resistant forms of bacteria develop in some cases, and toxic side-effects are not uncommon. Such treatment also prepares the ground for further disruption of the ecology of the natural flora, both of the genito-urinary tract and the gastro-intestinal tract, which are closely connected, resulting in more complication. 'Bladder cripples' can take heart as there is a safer and more efficient method of dealing with recurrent urinary tract infections — use of the friendly bacteria.

In numerous studies it has been shown, and is now generally accepted, that the source of the urinary tract infection is from bowel organisms. It may be of some comfort to women afflicted with 'frequency' and dysuria (discomfort, difficulty and pain relating to passing urine) that the condition usually does not involve the bladder but rather the urethra, the tube which carries the urine from the bladder to the outside world. Structural abnormalities (slight narrowing of the tube etc.) can allow bacteria a foothold in colonizing the region as the normal flushing of the tube may be affected, due to some local turbulence in the passage of the urine caused by the physical peculiarity. *Eschirichia coli* is the commonest pathogen to take advantage of this opportunity.

The urethra usually has resident bacteria, at the lower end. It is only when normal function alters that a spread occurs.

Treatment is best aimed at eradicating hostile organisms and also enhancing the natural defences of the region. A logical first step would therefore be to help to normalize the bowel flora from where the pathogenic organism spreads.

A noted Liverpool-based urologist, Dr R.M. Jameson, suggests that the regimen for anyone with 'frequency' and dysuria should involve strong dietary controls at the outset, involving three elements:

- A reduction in refined carbohydrate, via a sugar-free, high-fibre diet. This has the effect of altering the chemistry of the urinary tract as well as changing the ecology of the bowel flora, he maintains.
- Improved hygiene of the vagina and bowel — always wiping from front to back, for example, frequent use of bidets or douching with a hand-held shower, the regular boiling of any cloth used to wipe this region, the avoidance of foam baths, which can introduce bacteria into the bladder, use of natural fibre underwear etc.
- Topical applications (for example, creams containing iodine) which keep the region sterile.

Dr Jameson suggests the addition of *Lactobacillus acidophilus* or yogurt to the diet to assist in recolonization of the digestive tract. He states: 'It cannot be emphasized too strongly that the dietary regimen is designed to produce, not just symptomatic relief, but a life-long cure.'

Drs Fernandes, Shahani and Amer have also looked at the problem of urinary tract infections. They point to bacteria such as *Escherichia coli*, klebsiella, proteus and pseudomonas (as well as candida, see next chapter) as the major culprits. They explain that various strains of lactobacilli can adhere to and grow on the surface cells of the vagina and that: 'The normal urethral, vaginal and cervical flora of healthy females can competitively block the attachment of uropathogenic bacteria (*bacteria which cause disease in the genito-urinary tract*) to the sufaces of (genito-urinary) cells from women with and without a history of urinary tract infection.'

It seems that something in the wall of lactobacilli prevents the invading bacteria from being able to adhere to the vaginal and urethral surfaces, and that this fact can be used to help to prevent infection and to restore normality once infection has taken place.

A strategy should be adopted which includes both supplementation of lactobacilli (and bifidobacteria) as well as local applications to encourage repopulation of a healthy flora.

In addition, there is commonly an involvement of *Candida albicans* overgrowth in cases of urinary tract infection, and in the next chapter we will discover how the friendly bacteria can help to control both yeast and viral activity when they threaten the health of the body.

Chapter 8

Antiviral and antifungal powers

Just how lactobacilli can control virus activity is not certain. It may be by means of the changes in acidity which their prodigious production of lactic and other acids cause; it may be through the raising of the local temperature (viruses have poor tolerance of high temperature levels); it may be through the excretion or synthesis of other by-products (fatty acids perhaps), or it may be through a combination of all of these possibilities which are associated with their normal activities.

Revici's ideas

A remarkable Romanian physician who has practised in New York for the past fifty years (he is, at the time of writing, still in practice and well into his 90s) is Emanuel Revici MD. At his Institute of Applied Biology he has tackled fundamental problems of control of pathogenic organisms.

Behind his complex, and apparently remarkably effective, methods lies an easily understandable reasoning. He argues that, since in evolutionary terms the virus is older than the bacteria in its development, many forms of bacteria will have had to learn to 'deal' with and control viruses in order to protect themselves as they evolved. Similarly yeasts should have developed methods of keeping bacterial proliferation contained.

Such a hierarchy of development is seen throughout life in different areas of biological study. This suggests to Revici that we should look to the world of bacteria in order to find natural solutions to the problems of containing and controlling undesirable viral activity. In support of his hypothesis he has, through diligent research, identified essential fatty acid compound secretions, emanating from specific bacteria, which have viricidal effects (i.e. they kill viruses).

Some yeasts should also, according to Revici's primary concept, have powers of control over bacteria. We can see that this is true in the example of penicillin, a yeast, which is a powerful killer of bacteria.

Reversing the primary concept, it is argued that the 'older' organism (in evolutionary terms) can sometimes be seen to have learned to overcome such controls and to actually be able to damage or inhibit the more recent arrival. This we can observe in the case of *Candida albicans* proliferation, for example, in which some of the friendly bacteria can strongly inhibit yeast activity (to be discussed below).

Here there is an opportunity to use specific bacterial activity — acidophilus, for example — in order to control the rampant yeast, despite the yeasts having evolved after bacteria in evolutionary terms.

To summarize: In this miniature world of micro-organisms, teeming with life, battles and population surges, there have over billions of years evolved complex checks and balances. Biology is only just beginning to unravel some of these interactions, and Revici's concepts can help us to direct our minds and efforts towards the use of natural, biological, ecologically safe methods of treatment of viral and bacterial activity. This involves using other micro-organisms.

Acidolin and its antiviral potential

Research by Drs Hamdan and Mikolajcik in America examined the antibiotic properties of one of acidophilus's antibiotic products, acidolin, in relation to two deadly viruses. They exposed cultured colonies of polio virus and vaccinia virus to various concentrations of acidolin and observed the effects. In the case of polio virus, it was found that a dilution of one part of acidolin to 80 parts of inert liquid produced 'complete disintegration' of the viral cells. In the case of vaccinia virus, this disintegration was achieved at a concentration of one part in 160, showing this viral agent to be more sensitive to acidolin.

The researchers observed that as the concentration of acidolin increased, so did the acidity of the cell culture medium (from pH 6.5 at a concentration of 1 part in 320 to a very acid pH 3.6

when the concentration of acidolin reached one part in 80).

'At a high pH (6.5 to 6.8) no inhibition of the viruses was observed, indicating the possibility that inhibition may be due to the lowering of the pH in the cell culture system by the acidic nature of acidolin, and not due to specific antiviral factors associated with acidolin.'

In practical terms this need not concern us (i.e. whether the virus is deactivated by the acidity, or an antiviral substance, or both) since use of acidophilus culture either locally (such as on local virus-induced lesions in the mouth as described below) or in the GI tract itself would produce a lowering of pH and therefore a more acid medium.

Is this enough to deactivate viruses?

Herpes virus and acidophilus

In 1958 a Dr Don Weekes was treating several patients in Brookline, Massachusetts, who were suffering from the after-effects of diarrhoea which they had developed after courses of broad spectrum antibiotics. He was giving these patients active *Lactobacillus acidophilus* and bulgaricus tablets, and noticed that in each case there was not only an improvement in the diarrhoea problem, but also in other symptoms current, such as severe herpes simplex lesions in the mouth, as well as aphthous (mouth) ulcers (which are not caused by a virus but are more common when herpes virus is present and active).

The method of administration, for mouth lesions, was to dissolve four acidophilus and/or bulgaricus tablets (each weighing 0.25g) in the mouth with milk, 4 times daily. The milk was seen to be an 'activating' medium for the bacilli in the culture.

This experience led Dr Weekes to attempt to treat other forms of herpes virus infection such as herpes labialis (affecting the vaginal labia). In one study which he carried out, 64 patients with herpes labialis, 97 with aphthous stomatitis (mouth ulcers), 13 with dendritic ulcers and 6 with herpes infection of the genitalia, were treated by taking four tablets of *Lactobacillus acidophilus* and bulgaricus, with milk, 4 times daily.

The results were:

Herpes simplex labialis: 37 of 64 cured, 24 much improved, 3 no change. This is a success rate of 95 per cent which is quite

astonishingly good. The average time taken for the benefits to show was 3 days.
Aphthous stomatitis: 40 of 97 cured, 37 much improved, 20 no change. The success rate is around 80 per cent with benefits noted within 24 hours and lesions gone within 4 days for most patients.
Herpes progenitals: 6 of 6 cured, a 100 per cent success rate.
Dendritic ulceration: 6 of 13 cured, a 46 per cent success rate.

Only three of the patients treated noticed any side-effects, these being mild gastro-intestinal reactions.

Just how the lactobacilli achieved these results is not clear. Dr Weekes considered that it may result from a correction of previously low levels of salivary acid phosphatase. The raising in this way of the degree of acidity in the mouth might, he considers, be the factor which deactivates the viruses. He asserts: 'No claim is made that herpetic or aphthous lesion will not recur after cessation of therapy with viable lactobacilli. However, results indicate that individual attacks may be cured, improved or suppressed, by prompt use of the preparation. Used early enough it will actually abort the clinical aspects of the viral process, and will speed healing when used at any stage.'

He points to the preventive role lactobacilli might play when he describes a patient who developed aphthous ulceration of the mouth whenever she drank orange juice. By taking acidophilus daily she was again able to drink orange juice without the mouth ulceration. When she once more stopped the supplement the ulcers returned.

Remarkable results in Baltimore

Other doctors in the USA examined the efficacy of the supplementation of acidophilus/bulgaricus in the treatment of various types of mouth ulcers, and achieved outstanding results.

Again, a mixture of herpes-induced lesions and aphthous ulcers were treated using tablets which combined *Lactobacillus acidophilus* and bulgaricus (trade name Lactinex). Patients were randomly instructed to take from 2 to 4 tablets three times daily depending upon 'the severity of the disease, the physical size of the patients and the whim of the physician' for 2 to 3 days. Unless the patient reported 'unequivocal relief of the pain within 48 hours the results were considered to be negative'.

The results were in fact anything but negative. Of 40 patients treated (18 male, 22 female) 38 reported complete relief of pain within 48 hours, and 36 reported disappearance of lesions within five days. No side effects were noticed in this study.

The doctors conducting this study did not consider that they could give reasons for the efficacy of the method and we must return to the premise that it must be due to increased local acidity, or to some aspect of the secretions of these bacteria, or a combination of both.

Shahani's view

The name Dr Khem Shahani has by now cropped up a sufficient number of times for you to realize his enormous research contribution to our knowledge of the friendly bacteria.

On the subject of the antiviral capabilities of *some* lactobacilli he says: 'Selected and specifically grown strains of *L. acidophilus* have shown both antifungal and antiviral activity. Consequently, if manufactured under those conditions which augment the production of antifungal and antiviral constituents, acidophilus can retard the proliferation of vaginitis as well as flu or herpes.'

The question of quality

We will discuss quality control and the problems associated with poor commercial products in the shops in Chapter 16. On the heels of Dr Shahani's comment, above, a further observation of his at this time should act as a clear reminder that all is not well in the world of health food supplements.

'All acidophilus products available commercially are not prepared alike. The name on the bottle does not mean anything to the consumer if the bottle does not contain the right acidophilus strain. Many cannot even survive human gut-fluid and bile salts. Many products contain extremely low levels of living acidophilus cells, and those which are living are often unstable. Many manufacturers give the number of living cells at the time the product was formulated, or the bottle filled. But after the manufacture and storage the count goes down and the number of living cells can drop as low as zero.'

As we have suggested in previous chapters (and doubtless will do again) it is you, the consumer, who can most influence commercial methods by insisting when you purchase high potency cultured supplements, that they carry a guarantee of potency and that ideally this is linked to an expiry date.

Naturally, we would also want to remind you that the viability (its potential for survival inside you, coupled with its chances of forming active colonies of friendly bacteria) and the potency of the product, will be greater if it has not been centrifuged (this removes its supernatant and damages the structure of the cells and their colony-forming chains). This will be explained in more detail in Chapter 16.

The influence of commercial production methods on the effectiveness of different cultures should be kept in mind as we move from the control of viruses to the important problem of controlling the yeasts which live inside all of us, and which can at times become aggressively active, resulting in an amazing range of symptoms. The major culprit is *Candida albicans*.

Candida — the story so far

Moulds (fungi) are members of the plant kingdom, and yeasts are one subgroup of this family of organisms. Unlike other plants, they usually have no roots but absorb their nutrients from surrounding organic material by using enzymes which they secrete.

Moulds grow on most organic materials (something that is, or has been, alive); they also live in and on the soil, and they (or their spores) are numerous in the air, especially when it is humid.

A variety of these organisms have made the human body their home and, unlike the friendly bacteria, they offer no benefits in return. *Aspergillus* is the name of one which is associated with infections of the upper respiratory tract and the tonsils; *thrichophyton* is a yeast which causes 'athlete's foot'; and *Candida albicans* is a normal resident of the skin, lower bowel and vaginal regions. So, living inside all of us, are a host of uninvited squatters who have taken up residence in various desirable (for them) parts of our bodies.

The one which concerns us most at this time is the yeast with a sweet sounding name but some nasty habits, *Candida albicans*. This yeast is capable of causing conditions such as vaginal thrush (producing vaginitis) as well as oral thrush (yeast growing in the mouth) which is common in infants, as is 'cradle cap' another manifestation of candida's bad habits. Other far more serious complications can arise from its overactivity, as we will see.

Candida is present in every adult and in most children on the planet within a short time after birth. It normally does no harm, for it is kept in its place by our alert immune systems as well as by a variety of others of our micro-organism residents which (unlike candida) perform useful tasks in return for the living accommodation and food, which we provide. Among these is the group of friendly bacteria which we have been familiarizing ourselves with in previous chapters, including *Lactobacillus acidophilus*, bulgaricus and the bifidobacteria.

Recall that among the useful nutrients which these bacteria produce for us is one called biotin. This substance plays a very important role in preventing candida from turning into its rampantly aggressive mycelial or fungal form in which it puts down roots (rhizomes) and spreads rapidly. As Professor Jeffrey Bland explains: 'The conversion of the yeast form to the fungal form of candida is partially dependent on biotin deficiency. Japanese research has recently found that when biotin is added to the yeast in high levels it can prevent the conversion to its fungal form. They have also found that the fatty acid oleic acid (from olive oil) seems to prevent this same conversion.'

In its mycelial form candida's roots can penetrate the lining of the area in which it normally resides, such as the intestine. When this mucous membrane lining is penetrated by rhizomes it becomes possible for partially digested food particles, toxic wastes, and yeast breakdown products etc. to pass through into the bloodstream, resulting in allergies, sensitivities and constant drain on the immune (defence) mechanisms of the body.

Professor Bland explains: 'The yeast-like state [of candida] is a non-invasive, sugar-fermenting organism, whereas the fungal form produces rhizoids, or very long root-like structures, which can penetrate the mucosa, and it is invasive. Penetration of the gastrointestinal mucosa breaks down the boundary between the intestinal tract and the rest of the circulation allowing introduction into the bloodstream of many substances which are antigenic (i.e. stimulate the immune system to defend itself, possibly resulting in allergic reactions).'

Why does it change from a benign to an aggressive form?

Candida's change from simple yeast to mycelial fungus can be triggered by a variety of events which depress local or general immune function and the friendly bacteria which help to control the yeast. Among these is the use of antibiotics in the treatment of infection, for, as we have seen, not only are invading, hostile, micro-organisms killed by such drugs, but also friendly ones. If this happens to any great degree then one of the major controls of candida is removed.

Other events which can result in yeast overgrowth include use of steroid (hormone based) drugs such as cortisone and the contraceptive pill, or other drugs which are immunosuppressive. Ongoing infection by Epstein Barr virus or cytomegalovirus can also depress immune function and the body's ability to control candida. Also, a high level of sugar in the blood (such as occurs in a diabetic state or high sugar diet) is a predisposing factor to candida overgrowth.

What is the best way to know whether or not candida is active?

Regrettably, tests are not entirely accurate. One of the first doctors to pinpoint candida as a major cause of human suffering, Dr C. Orion Truss, says that the classic test for *Candida albicans*, a culture of a stool sample, does not always produce an accurate result, especially in a chronic infection. His experience is that a clinical trial using an anti-candida programme (see below) is best attempted when the general pattern of symptoms suggests candida as being in a state of overgrowth. If we rely on tests such as stool cultures, we may miss the diagnosis.

What about blood tests for antibodies?

We all have candida in and on us. The likelihood is that, in most people, some degree of antibody activity is present: this indicates that the immune system has defended itself against the yeast. However, if immune function is weak, such an immune response may be missing, and a test would show little or no antibody presence although candida continued to spread.

A far more reliable method of identifying candida activity is to look at your history and your symptoms. This can point quite accurately to a current candida overgrowth. In the list below we provide indications of the major symptoms associated with candida overgrowth. Candida is more likely to be active if one or more of the following has been, or is, true for you:

1. A course of antibiotics for 8 weeks or longer or for a number of shorter periods four or more times in one year.
2. Antibiotic treatment for acne for a month or more.
3. A course of cortisone, prednisone or ACTH (steroid treatment).
4. The use of contraceptive medication (the Pill) for a year or more.
5. Treatment with immuno-suppressive drugs.
6. More than one pregnancy.
7. Chronic multiple infections (viral, bacterial, parasites etc.) such as exist in anyone with AIDS or pre-AIDS syndrome (AIDS Related Complex). Rampant candida is one of the most obvious signs of a thoroughly depressed immune system, for whatever reason.
8. A very high sugar content in the diet.

The major symptom picture of active candida usually includes at least some of the following

If two or more of the symptoms shown on page 90 are common or current, and one or more of the list above is true for you, then candida is probably a current problem in your life, requiring specific action including, as a major feature, use of friendly bacterial supplementation in high doses.

1. Recurrent or persistent cystitis, prostatitis or vaginitis.
2. Endometriosis.
3. Thrush (oral or vaginal) more than once.
4. Athlete's foot, fungal nail or skin infection.
5. Extreme sensitivity to chemical fumes, perfumes, or tobacco smoke.
6. You feel worse after eating yeasty or sugary foods or drinks.
7. Multiple allergic symptoms.
8. Abdominal bloating, distension, diarrhoea or constipation, and/or itching rectum.
9. Premenstrual syndrome (fluid retention, irritability etc.).
10. Fatigue, lethargy, poor memory, inability to concentrate.
11. Muscular aches for no obvious reason, tingling, numbness etc.
12. Swollen or aching joints for no obvious reason.
13. Vaginal discharge, irritation, vaginitis.
14. Menstrual cramps or pain.
15. Impotence or loss of sexual desire.
16. Erratic vision; spots before the eyes.
17. Craving for sweet foods or alcohol.
18. Frequent upper respiratory tract infections, colds, post-nasal drip etc.
19. Depression.
20. Chronic acne.

What can we do about candida?

Any home wine maker will know that yeasts of all sorts love sugar. If the system is deprived, as far as it is possible, of sugar in concentrated or refined forms from food sources, this slows down candida activity. Other foods, such as chocolate, tea, coffee, vinegar and mushrooms are undesirable not because they 'feed' the yeast (except for their sugar content) but rather because they are all high in mould (e.g. on tea) or are derived from fermentation processes which involve yeast. These should also be stopped during the treatment of candida-related problems.

Because of the damage to the bowel mucosa and the spread of candida, the body easily becomes sensitized to yeasts, and is therefore more likely to react in an allergic manner to any substances or foods derived from or containing these (moulds and spores).

For this reason, in the first months of an anti-candida diet, any food or drink which has involved fermentation (other than live yogurt or foods based on lactobacillus fermentation such as kefir or acidophilus milk) or yeast in its production, or which is likely to have mould (old nuts, dried fruit etc.) on it should be avoided.

Anti-fungal substances which should be incorporated into an anti-candida programme should include:

- Olive oil (for its oleic acid)
- Garlic
- Biotin as a supplement to make up for the deficiency present in the GI tract, as explained above.
- Caprylic acid, a derivative of coconuts used as a fungus killer in preference to the commonly employed anti-fungal medical drug Nystatin, because the latter is itself yeast-based, and research at the Washington University School of Medicine shows that ultimately, after a period of treatment, when Nystatin is stopped it often results in even more colonies of yeast developing than were present before its use. Caprylic acid has no such rebound effect when its use ceases after candida is controlled. We never actually get rid of the yeast, remember, but only try to get it back under control.

Before we describe a suggested programme of diet and supplements for the control of candida, let us look at the evidence for the use of the friendly bacteria as a means of ensuring the most rapid control possible.

The lactobacilli and candida

Dr Shahani is quite clear on the usefulness of the lactobacilli against yeast infections. 'Ongoing research has revealed that supplementing the diet with friendly bacteria, like acidophilus and other compatible organisms like *Bifidobacteria bifidum* . . . should help in curing candidiasis.' This viewpoint is confirmed by Japanese research (1984) which examined the degree of overgrowth of candida as expressed by the levels found in the faeces of patients with leukaemia who were receiving drug therapy. The candida counts were very high indeed before treatement with bifidobacteria which reduced the levels of candida in the faeces of some patients from a high of 100,000,000 per gram to a mere 10,000 per gram after treatment. The effectiveness of bifidobacteria in achieving this was seen in all 16 patients treated, whereas 11 'control' patients not receiving bifidobacteria supplementation showed no change at all in their candida levels.

Many researchers report that *L. acidophilus* seems to produce substances which retard the growth of candida, and this is borne out when *L. acidophilus* is added to culture dishes in which Candida is growing, where an ability is seen to slow and even stop its growth. An additional bonus is received when bifidobacteria are supplemented against candida, as this has a uniquely powerful ability to enhance detoxification via the liver, as well as its now familiar role in the GI tract itself.

For these many reasons we can only echo the words of Professor Jeffrey Bland when he states: 'We have been very excited about an alternative therapy for the management of candida infection, which avoids the use of anti-yeast medication. It is well recognized that a disturbed flora of the GI tract can establish a proper environment for yeast proliferation. By reinocculating the bowel with the proper symbiotic acid-producing bacteria, there is a reduction in the compatibility of the intestinal environment for the yeast proliferation. We have recently used an oral supplement of *Lactobacillus acidophilus* . . . this has been extremely successful in reducing *Candida albicans* in the intestinal tract. The *L. acidophilus* is given as a dry culture.'

This is seen to play a major part in the strategy which we outline below as being suitable for anyone with active candida overgrowth.

The anti-candida diet and supplementation programme

Anyone with active candida overgrowth should avoid, at least during the first few months, foods based on fermentation (wines, vinegar, miso etc.) or which are yeast-based (mushrooms, yeast extract, spreads etc.) or which might be expected to contain high levels of mould, spores or fungi (cheese, dried fruit, tea etc.). These should all be avoided for a period of several months, until the candida activity is well under control and an element of desensitization has taken place through lack of contact with these substances.

In the early stages (first 3 to 4 weeks) of following this diet we suggest that anyone with an obvious candida problem should not eat any fruit and, more importantly, should not drink any fruit juice. This is because of the relatively high natural sugar content of fruits and the rapid absorption of these sugars when the fruit is juiced. This restriction may be relaxed after the first month with the gradual introduction of fruits, although melons should continue to be avoided until recovery is well under way.

Avoid the following:

- All sugar, whether white, brown or shades between. This includes honey, sweets, glace fruits, jellies, jams, pickles, sauces, preserved fruits and anything else which has 'hidden' sugars.
- All alcohol of any sort.
- Anything made from refined (white) flour products (buns, cakes, pastries, crackers, biscuits, pasta etc.). These can be safely and beneficially replaced with foods made from wholegrain cereals (brown pasta, wholegrain bread etc.).
- Any foods or drinks containing colouring, flavouring or preservatives.
- Frozen vegetables which contain sugar (peas and many others).
- Breakfast cereals containing sugars or made with refined flours.
- Fruit juice (unless diluted 50/50 with water and sipped slowly).

- Stimulants which trigger the release of sugar in the body, such as tea, coffee, cola and cigarettes.
- Any meat or fish dishes which have been cured, smoked, preserved, pickled or in any way processed. This includes sausages, salami and bacon.
- Fruit (for the first month or so).
- All yeast-derived or fermented foods including: bread (apart from non-yeasted varieties), mushrooms, soya sauce, Indian or China tea, nuts and seeds (unless fresh and free of rancidity and mould), cider, beer, ginger ale, wine etc., cheese (apart from cottage cheese), all malted products, vinegar of any sort, saurkraut, pickles, relishes and sauces, anything containing monosodium glutamate (rich in Chinese restaurant foods), anything cooked in bread-crumbs, anything containing citric acid (usually yeast derived) and, for the same reason, most canned or frozen citrus drinks, mayonnaise.
- Multi-vitamin/multi-mineral/B-complex vitamins/individual B vitamins/selenium supplements, *unless guaranteed from a yeast free source*.
- Antibiotics should be avoided by anyone with candida unless absolutely unavoidable.
- Meats and poultry which might be from sources where antibiotics are used should also be avoided.

Rotation diet? An additional dietary strategy is worthy of being considered in which anyone with continuing symptoms of food sensitivity or allergy (all too common in people with candida) can learn to 'rotate' the eating of problem foods or food families (for example, all grains or all dairy produce).

This strategy of planning a weekly menu in accordance with your known (and sometimes unknown) problem foods may require the help of a skilled expert in nutrition, although books are available to guide the self-help application of the concept. It has been found that if problem foods are rotated, so that they do not appear in the diet more than once every four days or so, the reaction to them is lessened or disappears altogether. This in itself reduces the constant drain on immune function which repetitive allergic responses produce.

A word about supplements

The supplements which are suggested below are designed to perform several tasks. Most are specifically immune system enhancers, whereas others have different roles to play (aiding digestion, killing candida, speeding detoxification, enhancing protein supply etc.).

Some have anti-candida actions, whereas others enhance the way the body metabolizes and deals with sugars (chromium, for example).

There are a lot of supplements to be taken. This may require some getting used to, and we suggest that you look on these not as medicines but as the concentrated foods which they are. If candida is rampant we cannot stress strongly enough the importance of avoiding yeast-based supplements, and this is possible by purchasing specific brands of nutrients which guarantee that they are not extracted from or derived from fermentation or yeast.

Some nutrients, such as the amino acids (from which protein is made) and enzymes, need to be taken on their own with water away from meal times. Other nutrients should be taken at mealtimes (during or after) unless specific instructions to the contrary are given. In most instances divided doses are better than just one large dose. Such advice is given in the text with the dosage suggestions.

It is a great help to map out a plan of action relating to the taking of supplements in which the supplement, its dosage and any special instructions (with water etc.) are written down clearly. The timing of the taking of each supplement should then be listed in relation to mealtimes etc. (some are taken on waking, others between meals, others with food, some after meals, others at bedtime etc.). In this way a clear plan will emerge, and in a short time it will become second nature to follow it. Trying to remember all the variables involved in such a programme without a planned, written, easily accessible guide, is usually doomed to become haphazard and ineffectual.

Enough protein?

If candida overgrowth and allergic reactions are concurrently in evidence, then the pattern of eating could be very restricted indeed. It is of the greatest importance that an adequate protein intake be assured. This may present a problem for anyone with a compromised digestive system, and we therefore recommend the use of what are called 'free form' amino acids in supplemental form.

We all need to find in our food each day a selection of essential amino acids from which we can make all the other amino acids, and from this raw material restore and build the tissues and cells of the body. There are 8 (some say 10) essential amino acids which have to be present in our food and which, in order to be of any use to us, need to be adequately digested. Unless these are separated from each other (they arrive in food in chains) so that they are 'free', we cannot use them to construct new tissues. Because many people with dysbiosis or candida have inadequate digestive function, it is useful to take ready-prepared amino acids which the body can then alter to make up the complete set of 20 or so amino acids which form part of everything in the body.

Between 5g and 15g daily of free form amino acids (the whole complex, ideally, or just the essential ones) should be taken half an hour before or an hour and a half after meals, with water.

Other digestive aids

If digestion is poor, the addition of enzymes to aid in digestion of the various foods eaten can be a major help. For assistance in protein digestion take proteolytic (protein digesting) enzymes such as *Bromelaine* and *papain*.

If dyspepsia is regularly felt after eating there may be a shortage of hydrochloric acid, and this too can be supplemented (in Europe, at least, but no longer in the US because of FDA regulations). Health food stores or pharmacists supply betaine hydrochloride and pepsin in tablet form for use at the beginning of meals. In the US it is suggested that advice be sought for any such digestive problem from a health professional who is an authority on nutrition. The regular use of high potency friendly bacteria will itself help in such problems.

Anti-candida supplementation

The specific anti-candida nutrients include:

- *High potency acidophilus* powder. A vital part of the anti-candida programme. This must be a viable, active form (which would therefore require refrigeration at all times) together with other lactobacilli such as *Lactobacillus bulgaricus* and bifidobacteria supplements. Dosage for someone with candida is between a half and a whole teaspoonful of *L. acidophilus* powder together with a half to a whole teaspoonful of bifidobacteria powder in filtered or spring water, two or three times daily *away from mealtime*; and a quarter to a half teaspoonful of *L. bulgaricus* in water, juice, milk or cooled foods, with meals. Additional aid to the digestive tract can be achieved by consuming live yogurt which contains symbiotic friendly bacteria such as *S. thermophilus*.

- *Biotin*. A B-vitamin in 500 microgramme doses with food.
- *High potency garlic* capsules (unless a great deal of raw garlic is being eaten) for its anti-fungal activity. Take 3 to 6 capsules daily with meals.
- *Oleic acid*, which is simply the active portion of olive oil, is also anti-fungal. A dessertspoonful (a tablespoon in America) of oil daily on salad.
- *Caprylic acid*. A non-nutrient extract of coconuts which is a powerful anti-fungal agent and completely non-toxic (as it is not absorbed, but simply passed through the gut, killing yeast). This is obtainable without prescription from health food stores or health professionals. Three time-release capsules containing caprylic acid should be taken with each meal.
- *Botanical seed extract (grapefruit seed)*. Used in America as an alternative to caprylic acid.

Note: During periods of rapid yeast destruction the body is called on to detoxify the breakdown products of this process, and this can lead to your feeling particularly seedy, nauseated and off-colour. The reaction is known as yeast 'die-off' or 'burn-off' and can last for some days or even weeks. The use of proven strains of high potency bifido-bacteria and the general dietary strategies discussed above should minimize this. Do not, however, stop the anti-candida programme if such a feeling begins, as this is a critical stage of the treatment which, if stopped suddenly, can lead to a rebound of the candida and even greater feelings of ill-health.

- *Germanium*. This nutrient is a recently researched one with energy enhancing capabilities. It also has powerful anti-fungal properties and its supplementation is recommended at a dosage of not less than 100mg (and up to 300mg) daily. It is expensive, however, and will continue to be until a cheaper method of production is developed.
- *Aloe vera juice*. This is derived from the desert plant and it is an anti-fungal substance. Several teaspoonsful in water should be consumed daily

These suggested therapeutic substances and nutrients should be used for not less than six months by anyone with an ongoing candida problem.

Direct acidophilus application for vaginitis and rectal itch

If candida overgrowth is producing, or contributing to, symptoms such as vaginitis, a useful strategy is to apply high potency cultures directly to the region affected. This has the effect of increasing local acidity (lowering pH) as well as killing yeast cultures in the region. Such measures do not replace the systemic approach to dealing with candida, but actively support it, as well as providing dramatic symptomatic relief from burning, itching and irritation.

Some firms supply special disposable applicators. These have prelubricated soft tips, and are designed to be used in order to insert such cultures after they have been mixed with a small quantity of yogurt.

First prepare the mixture of, say, a quarter to a half teaspoon of a viable high potency acidophilus powder in enough yogurt to fill the applicator.

For Vaginal use: Lie on your back, knees bent, and insert filled applicator gently into the vagina as far as is comfortable. Squeeze the applicator firmly and constantly until it is empty, remove and discard.

For Rectal use: Lie comfortably on one side with the upper knee drawn up to your chest (or face down with a pillow under the hips) and slowly insert the filled applicator completely (as far as is comfortable) and empty its contents by squeezing firmly until empty. Remove applicator and discard.

The use of a sanitary napkin can ensure that no leakage occurs, although this is seldom a problem.

Note: If no other aspect of the anti-candida programme can be followed we urge you to, at the very least, introduce the supplementation recommended of the friendly bacteria, see the notes on the dreadful candida/staphylococcus aureus symbiosis on the next pages, if you need more motivation!

Professional advice

In the USA we recommend: Dr Robert J. Marshall and Associates, Torrance, California, whom we feel has the most direct clinical experience in the use of high quality probiotics in the US.

In the UK we recommend people with candida problems to contact: The Candida Albicans Advice Group, PO Box 89, East Grinstead, West Sussex RH19 1YY. (This is a referral service which provides information and directs enquirers to one of a network of doctors and other health professionals specializing in candidiasis.)

The deadly connection: *Staphylococcus aureus* and *Candida albicans*

Before we reveal research evidence of a terrible symbiosis between *Staphylococcus aureus* and *Candida albicans* (both of which can be controlled by the friendly bacteria) let us briefly examine some of the proven disease states which *S. aureus* is known to produce.

Toxic Shock Syndrome (TSS): Typically a young woman (95 per cent of cases) affected by TSS begins to notice symptoms on about the fifth day of her period. She will probably be using tampons. Her symptoms include a widespread rash, fever, watery diarrhoea, vomiting, sore throat, headaches and aching muscles. The skin may begin to peel off, kidney failure may ensue and both respiratory and cardiac complications are common. One or two out of every ten people affected by TSS will die.

The cause? Rampant infestation with *Staphylococcus aureus*, with the symptoms being the result of toxins which this bacteria produces.

Scalded Skin Syndrome (SSS): This condition, which affects infants, young children and immune-suppressed adults, usually follows a spread from a primary infection elsewhere in the body, such as conjunctivitis. SSS is characterized by fever and profound weakness and a bright red, very tender, skin rash involving large blisters which slough off in sheets of skin, leaving large areas of the body without skin at all. Even normal looking skin shears away with light pressure. Complications occur relating to the body's temperature control and fluid balance mechanisms being disrupted in this way. Again, as with toxic shock syndrome, this is caused by a toxin secreted by the bacteria, *S. aureus*.

Between 80 and 90 per cent of *S. aureus* infections which are found in hospital settings are resistant to penicillin treatment.

Other conditions associated with *S. aureus* include gastroenteritis, bone and joint infections (osteomyelitis) and septic arthritis, pneumonia, meningitis, inflammatory heart disease (endocarditis) etc.

The connection: In a series of experiments on animals, Dr Eunice Carlson, of Michigan University, established that when there was a combined infection of *S. aureus* (or *Streptococcus faecalis* or *Serratia marcescens*) and *Candida albicans* there was an enormous potentiation of the infection caused by the bacteria.

She tells us: 'Although these studies show that candida has a strong amplifying effect on the virulence of other organisms (*S. aureus*., *S. faecalis*) how this is achieved is a mystery.' She continues: 'One possibility is that the candidal infection process causes physical damage to the organ walls which makes them 'leaky' allowing other microbes or chemicals (perhaps toxins) or both to penetrate more easily: it is also possible that candida directly stimulates the growth of *S. aureus*.'

Why, therefore, don't doctors treat candida aggressively

in conditions such as those decribed? Because they either do not know it is a part of the problem or they cannot 'prove' its presence.

Dr Carlson discusses this conundrum, and gives the example of the known infestation of dentures with candida. 'It is now believed to be very common and to occur in 60 per cent of all denture wearers. Biopsies of inflamed areas, however, consistently fail to demonstrate tissue invasion [with candida]. We can speculate that an equivalent infection of the small intestine would be virtually undetectable.

As we have seen in Chapter 8 we have to assume candida's active presence on the basis of symptoms, not scientific tests, which are usually inconclusive.

Dr Carlson continues: 'Physicians have reported therapeutic cures for a variety of diverse disease conditions by anticandidal drugs. *It now appears possible that this fungus may play a key role in many disease conditions, not by its own toxic or invasive growth, but rather by enhancing secondary infection*.'

We now have the knowledge, together with the availability, of high potency forms of those friendly bacteria which have proven abilities to inhibit and control both *S. aureus* and *candida* (see Chapter 8). We should use them.

Chapter 9

The cholesterol fighting potentials of the friendly bacteria

Cholesterol is found in every cell membrane in your body, and without it normal metabolic processes would be impossible, but an excessive level of cholesterol in the blood (hypercholesterolaemia) is considered to be one of the main factors which predisposes towards hardening of the arteries and coronary disease, the number one killer of modern times in Western industrialized countries.

In atherosclerosis the plaques which accumulate on the walls of arteries are largely made up of cholesterol which circulates in the blood as part of particles called low density lipoprotein (LDL). The higher the levels of LDL in the bloodstream the faster atherosclerosis develops. Another form of lipoprotein, a high density version (HDL), is found to be in plentiful supply in people with low levels of cholesterol, low incidence of atherosclerosis and, therefore, of good cardiac health. HDL seems to play an important part in actually removing excessive levels of cholesterol from tissues. The term used to describe excessive levels of cholesterol in the blood is hypercholesterolaemia.

It is not generally realized that the body manufactures far more cholesterol itself than is contributed to it in the diet. This manufacturing process is normally directly tied to the requirements of the body, with more being produced when the diet is rich in fats, especially those of animal origin (dairy foods, meat etc.).

The various bile acids are manufactured (conjugated is the technical word for this) in the liver from basic raw material such as cholesterol. In the liver, cholesterol is turned into cholic acid and chenodeoxycholic acid which are combined with the amino acids (building blocks of protein) glycine and taurine.

Once in this conjugated form they are stored in the gall-bladder, from where they are passed to the duodenum after a meal. The bile salts help the digestive process by dispersing fat globules, making them easier to metabolize and digest, acting as efficient detergents (which they are). After being used in this manner the bile salts are mainly reabsorbed from the small intestine and are returned to the liver for re-use.

Before this can happen, however, the bile salts need to be broken down into their basic constituents again. The bifidobacteria (and others) have the power to deconjugate (literally take apart) the bile salts back into their original constituents for this process of recycling or elimination (in the faeces). If there are excessive quantities of bile salts (as in a fat-rich diet) the bifidobacteria (as well as numerous other bacterial inhabitants of the colon) are inhibited or damaged. This is one more good reason for avoiding a high fat intake.

The amazing Masai

The Masai tribesmen of Africa live a nomadic existence, herding their cows from region to region, and virtually living entirely on a diet derived from them. They seldom eat the animals, however. Rather, they drink fresh blood from the living beast mixed with fermented milk from the same source. Their intake of saturated fats is enormous, and yet their levels of cholesterol are well below those found in residents of New York and Paris, and they have extremely low levels of heart disease. Why should this be the case?

Research into this question resulted in the discovery that there must be an 'anti-cholesterolaemic milk factor' (AMF) in the fermented milk, which had the ability to reduce the levels of cholesterol. Much effort has gone into trying to identify just what AMF might be, and thus far there appear to be two contenders: hydroxymethylglutaric acid and orotic acid. Whichever it is seems to inhibit an enzyme which synthesizes cholesterol (i.e. puts it together from other elements). Similar anticholesterolaemic effects have now been demonstrated from constituents of live yogurt.

Fermentation does not seem to be an essential factor in producing the AMF effect as there are studies which show it to be present in unfermented sweet acidophilus milk. The degree of concentration required for this effect, according to Dr Shahani, was of 4 million viable acidophilus organisms (Strain DDS-1) per millilitre of milk.

Research

Both animal and human studies now show that cultured milk products such as yogurt and milk, to which acidophilus cultures have been added, efficiently lower levels of cholesterol in the blood. For example, the feeding of rats with fermented milk or milk containing active *Lactobacillus acidophilus* decreased cholesterol levels in the rats' bloodstreams. A similar reduction was achieved in pigs, previously kept 'germ-free' and then fed *Lactobacillus acidophilus* in order for it to colonize their GI tracts.

Rabbits fed on yogurt showed a marked reduction in blood cholesterol levels, and the same thing happened when hens were fed *Lactobacillus acidophilus*.

In some studies, however, no such benefits were observed and a conclusion was reached that, as with the experience relating to their antibiotic production, the ability to reduce cholesterol levels relates directly to specific strains of acidophilus, although as yet insufficient evidence exists as to just what these variations are.

The *Lactobacillus acidophilus* strain RP32 was able to deconjugate cholesterol easily (in an oxygen-free environment such as occurs in the large intestine) in the study which involved pigs, whereas other strain, P47, could not do so. Much research remains to be done.

What do the bacteria do to the cholesterol?

Bacteria in the GI tract have the ability to metabolize (break down or alter) both hormonal products (oestrogen etc.) and bile salts. When animals are kept 'germ-free' with no colonization of their GI tracts at all the bile remains relatively unchanged from the original version produced for digestion (the conjugated form of bile).

When suitable bacterial cultures, such as the bifidobacteria and *Lactobacillus acidophilus*, occupy the GI tract the bile exits the GI tract (either back to the liver or out of the body in faeces) in various forms, conjugated, deconjugated and with various breakdown products of bile acids. It is in these changes of the bile acids that cholesterol is somehow removed from the systems of these experimental animals.

Laboratory studies

Dr S. Gilliland and his colleagues at Oklahoma State University acknowledge that trials in both humans and animals proved that there was a decrease in levels of cholesterol in the bloodstream after the ingestion of *Lactobacillus acidophilus*; but they also note that no-one had yet shown that this was because acidophilus acted directly on the cholesterol. They therefore conducted trials which showed that some strains of *Lactobacillus acidophilus* actually have the ability to assimilate cholesterol.

This was shown in cell cultures in laboratory conditions, when acidophilus was grown in the presence of cholesterol. *Some cholesterol actually appeared inside the lactobacillus cells while they were growing, and this was associated with a reduction in the amount of cholesterol in the surrounding growth medium*. This only happened in conditions where there was no oxygen and when bile was also present, precisely the conditions which would be found in the intestinal tract. Dr Gilliland stresses that the amounts of cholesterol placed in the culture dishes was not in excess of the quantities which would normally be found in the intestines.

Their conclusion was that this ability of *Lactobacillus acidophilus*, 'Would make it possible for the organism to assimilate at least part of the cholesterol ingested in the diet, making it unavailable for absorption into the blood.' A similar action could be expected on cholesterol manufactured in the body, if it reached the intestine. We now have evidence of two ways in which the friendly bacteria help to reduce cholesterol, by direct absorption and by deconjugation.

What about humans?

Animal and laboratory studies can only take us so far. We need to have evidence of humans showing reduced cholesterol levels when using these cultures in their diet, in order to be able to authoritatively suggest such use of friendly bacteria in the quest for better heart health. A 1979 study attempted to provide this evidence. It involved 54 volunteers aged between 21 and 55

(24 males, 30 females) all in good general health with no history of cardiovascular disease or gall bladder disease, who were studied for 12 weeks, during the first and last four weeks of which they were supplemented with either unpasteurized yogurt or milk.

In a separate study volunteers were supplemented for the full 12 weeks with either unpasteurized yogurt, pasteurized yogurt or milk. Both forms of yogurt contained live cultures of *Lactobacillus bulgaricus* and *Streptococcus thermophilus*.

(Note: Please refer to Chapter 3 in which we briefly explain the variations in use of pasteurization. Yogurt which is pasturized after it is cultured results in reduction of viable organisms. This may not greatly influence the effects of yogurt on cholesterol — see below — but would certainly detract from other functions of the friendly bacteria.)

No other changes in the diet or behaviour, such as smoking habits (12 were smokers), or exercise, were suggested during any stage of this research. A great deal of detailed information was taken during these studies relating to blood constituents, as well as precise data as to nutritional patterns.

The major finding was that yogurt supplementation (pasteurized or unpasteurized) causes a significant reduction in blood cholesterol levels. It is important, we believe, that during the studies there was no significant alteration in the total calorie intake, or in the amount of cholesterol, protein, carbohydrate, fat or fibre consumed, indicating clearly that the effect of the yogurt in reducing cholesterol levels was independent of all other influences from the diet.

The effect of decreased cholesterol levels was seen to start within seven days of the beginning of yogurt intake, by which time a reduction of between 5 and 10 per cent in cholesterol levels was observed. The beneficial effects stopped by the fourth week after ceasing yogurt intake.

The doctors who conducted this study refer to other similar research programmes, one of which indicated that when yogurt was supplemented in large amounts (enough to make up nearly half of the daily calorie intake) there was a substantial reduction in cholesterol levels amongst the healthy volunteers. This study also showed that the synthesis of cholesterol in the body was interfered with during this heavy course of yogurt supplementation, indicating one way in which the benefits were being achieved.

So how is cholesterol controlled by the friendly bacteria?

The researchers state that this reduction could be because of alteration in cholesterol synthesis, absorption, conversion into bile acids or the synthesis and degradation (breakdown) of lipoprotein. No firm conclusions have been reached.

Naturopathic doctor Carl Hangee-Bauer, suggests: 'There are a variety of mechanisms to explain the apparent hypocholesterolaemic results of lactobacillus fermentation. Lactobacilli may metabolize cholesterol for energy, suggesting that cholesterol may be a 'food' for this organism. Other possibilities include alteration in bile acid metabolism due to modification of the bowel flora, which would favour increased excretion of cholesterol (and other factors)'.

In fact, it is not only as a means of obtaining energy that friendly bacteria act on bile salts. Dr M. Speck of North Carolina State University, has shown that when the bile is deconjugated (broken into its constituent acids) the resulting free acids have an inhibitory effect on the growth and activity of rival organisms. Remember that as a result of the breakdown of bile salts, which leads to lowering of cholesterol levels, the normal passage of certain of these acids in the circulation between the liver and the GI tract is maintained.

Once again we see that the friendly bacteria, whether they are trying to colonize us, find food, or protect themselves, always seem to provide a series of very useful by-products in terms of our health.

Chapter 10

The anti-carcinogenic and anti-tumour potentials of the friendly bacteria

The World Health Organisation has estimated that a minimum of 60 per cent of the cancers in women, and almost half those in men, can be directly related to diet and nutrition. If we add to those the many which are related to personal abuse (smoking, toxin exposure etc.) an enlightening truth is revealed to us, that the vast majority of cancers are due to factors over which we are, or should be, able to exercise control.

Those forms of cancer most directly related to our diet include those of the following organs:

Organ	Major carcinogenic factors	Major protectors*
Oesophagus	tobacco and alcohol	stopping smoking and drinking alcohol; antioxidant nutrients (vitamins A, C, E etc.)
stomach	nitrites in our foods or formed in the GI tract	Antioxidant nutrients
Colon	fats, fried foods, hormones, nitrites	fibre, vitamins, less fat
breast/prostate	fats, fried foods, hormones	decreased fats and fried foods
pancreas	fried foods, tobacco	decreased tobacco, fat and fried food

* And, of course, as will become clear below, the lactobacilli.

In general the effects of deficiencies of essential nutrients, or the presence in the body of toxic substances, have an indirect effect on the development of cancer, rather than there being a specific cause and effect relationship.

One of the key reasons for the uncontrolled proliferation of cells, which characterizes cancer, is the formation, or presence, in the body of chemicals which have the ability to cause the cell changes which result in cancer. These are known as carcinogenic factors.

Diets which are high in animal protein and fat seem to be the ones which most add to the risks of some forms of cancer, most notably of the breast, prostate and pancreas, as well as the stomach and colon. The changes which a fat-rich and protein-rich diet produce in the numbers and behaviour of the microflora of the GI tract seems to increase this susceptibility to cancer developing.

Research points to chemical changes engineered by enzymes produced by bacteria (other than the lactobacilli) in the GI tract, as being very much involved in the formation of carcinogens, when the diet is unbalanced in this way.

Conversely, some of the friendly bacteria seem to have the ability to inhibit the formation of carcinogenic chemicals by 'switching off' the enzymes produced by the harmful bacteria involved in this process. Evidence exists in abundance that populations which consume relatively high quantities of cultured milk-products have lower levels of cancer incidence, as a result of just this protective behaviour of the friendly bacteria.

One major danger of dysbiosis (disturbance of the ecology of the bowel) is that it results in the production in the bowel of a large number of highly toxic and sometimes cancer-causing substances. When the diet is rich in meat and fat, many of the potentially hostile bacteria are encouraged to produce certain enzymes. They have the ability to convert certain relatively harmless chemical by-products of the process of digestion (called procarcinogens) into extremely harmful carcinogens. Research on human subjects has shown that many of these enzymes (examples include b-glucuronidase, b-glucosidase and nitro-reductase) all have a significantly lowered degree of activity when *Lactobacillus acidophilus* is supplemented.

Drs Fernandes, Shahani and Amer have classified the three ways in which the friendly bacteria display their anti-carcinogenic properties:

1. They eliminate substances which can become carcinogenic (i.e. which are procarcinogens). Among these are chemicals called nitrites which are used in food processing and which may then be converted into nitrosamines in the gut. Nitrosamines are carcinogenic substances. The lactobacilli have the ability to deactivate nitrites before they can be converted to nitrosamines and also to metabolize the nitrosamines themselves back into non-carcinogenic substances. As in many other examples we have seen, some strains of lactobacilli are more efficient at these potentially life-saving functions than others.
2. They alter the enzymes (such as b-glucuronidase and nitro-reductase) produced by other bacteria in the GI tract (e.g. bacteroides, clostridium etc.) which have the ability to convert procarcinogens (such as nitrites) into carcinogens (such as nitrosamines). In general the more of these enzymes there are, the greater the risk of cancer forming substances being created. The ability of lactobacilli to deactivate production of (or the action of) these enzymes is the most important contribution to cancer prevention. We have, therefore, seen that a twin potential is offered by the friendly bacteria: they reduce the quantities of potential carcinogens and they also inhibit the activities of those bacterial enzymes which turn potential carcinogens into actual carcinogens.
3. They appear to have the ability to actively suppress some tumour activity.

Enhanced immune function as well?

A further possibility, over and above these different potential benefits, all of which will be examined below, is that aspects of immune function may be enhanced (potentiated is the scientific term) by means of supplementation in the diet of specific friendly bacteria.

The immune system is a vast inter-related complex of defensive capabilities. It can, for ease of understanding, and for reasons of a natural division of activities, be separated into either what is called *cell-mediated* protective immunity — which is directed at achieving several tasks including defence against viruses, fungi and many bacterial invasions; rejection of tumours and transplanted tissues — or *humoral immunity* — which involves those defence capabilities which are mobilized in the blood, including antibodies which combat viruses and bacteria and immediate allergic type responses.

One of the most direct 'weapon systems' we have, as part of our cell-mediated immunity, involves the activities of cells called macrophages. These literally 'eat' enemy cells, be they tumour cells or invading micro-organisms, in a process called phagocytosis (Greek *phagos* — to eat).

Enhancing mouse immune systems

In studies carried out in Argentina in 1987, mice were examined in order to assess their immune system's response to the regular eating of food containing *Lactobacillus acidophilus* and *Streptococcus thermophilus*. The mice were divided into groups, some of which received, in addition to their normal diet, either live or deactivated yogurt bacteria in their food. Other mice were injected with either active or deactivated cultures of these bacteria. A further group of mice were untreated in any way, and were kept merely as controls to compare levels of macrophage (and other) activity with those mice being supplemented.

From the second day onwards it was found that the mice supplemented with acidophilus (either by mouth or injection) showed between three and four times as much macrophage activity as control mice. Those mice receiving *S. thermophilus* (a transient yogurt forming bacteria, not a resident of the GI tract as is acidophilus) showed some increased macrophage activity in response (but not as much as was seen in those mice getting acidophilus).

The researchers point out that: 'Since activation in the body of macrophages is important in suppressing tumour growth, immuno-stimulation by the oral route might well be a new approach by stimulating the specific and non-specific immunity of the host.'

Put simply, there are advantages to the immune system, in terms of prevention of tumour growth or development, to be gained from the regular intake of these lactobacilli, which enhance those body cells which are specifically related to the destruction of tumour cells. There is no reason at all to believe that human immune response to acidophilus or thermophilus supplementation would be any different, and some of the research described below supports this view.

Animal studies relating to prevention of cancer

We wish to make it clear that while we do not advocate animal experiments, for a variety of reasons, it is obvious that some important evidence which is of use to man has been uncovered in this manner, and we feel bound to share such knowledge with you.

When rats were given the cancer causing DMH (1,2,dimethylhydrazine) while they were living on an all grain diet, just 31 per cent of them developed cancer of the colon. When given the same toxin whilst living on a beef-rich (and therefore fat-rich) diet, a full 83 per cent of the rats developed colon cancer. When given *Lactobacillus acidophilus*, as well as the toxin and the beef diet, only 40 per cent developed colon cancer after 20 weeks. By the time 36 weeks has passed 77 per cent were afflicted.

In this unpleasant experiment we see clear evidence of the protective role played by acidophilus in the face of a known cancer causing agent, DMH. We can also see just how much the changed diet (going from grains to beef) caused the microflora to allow the carcinogenic toxins to act in a far more lethal manner.

Human studies relating to prevention of cancer

A. The way friendly bacteria eliminate substances which can become carcinogenic

During the digestive process, enzymes secreted by certain bacteria encourage the formation of cancer-causing nitrosamines from food constituents such as nitrites. The manufacture of these carcinogens often involves the activities of putrefactive bacteria, found in active abundance in the GI tracts of people who eat a great deal of meat and meat-fat (see below). Two of these nitrosamines, dimethylamine, or DMA, and dimethylnitrosamine, or DMN, can be formed in the stomach from nitrites in the food.

A great deal is now known about how these and other amines can be formed by bacterial action, whereas relatively little research has been done in trying to find out how other bacteria can deactivate these dangerous nitrosamines.

A research study in England by Rowland and Grasso attempted to discover which of the common bacteria in the GI tracts of mammals (including humans) could achieve this important detoxification process. The carcinogenic nitrosamines which were examined in this way were DMN and another called diphenylnitrosamine, or DPA.

Most of the bacteria tested were able to some degree to degrade DMN and DPA, including *E. coli*, bacteroides, bifidobacteria, lactobacilli and *Streptococcus faecalis*. The bacteria which were found to be the most efficient at degrading DMN and DPA back into safe chemical elements were *E. coli* and the lactobacilli.

Bifidobacteria were able to achieve this effect, but far more slowly. The nitrosamines tested showed vast differences in the speed at which they could be 'made safe', with DMN taking five times as long to degrade as DPA.

B. The way the friendly bacteria alter the production of enzymes which produce carcinogens from procarcinogens

1. Drs Friend and Shahani wanted to see what effect dietary supplementation of unfermented *Lactobacillus acidophilus* milk would have in humans, particularly in relation to some of the known enzymes which act to encourage the development of cancer in humans, by turning procarcinogens into carcinogens.

They therefore looked for changes which might take place in the microflora of two groups of geriatric patients who had their diets supplemented with either plain milk or acidophilus milk. The milk addition made no difference to the microflora. The acidophilus milk, however, resulted in a reduction in the activity of two of those enzymes which produce undesirable changes, b-glucuronidase and b-glucosidase.

The very high levels of lactobacilli which supplementation succeeded in establishing in the GI tracts of these patients, and with this colonization their protective effects, was found to continue to be measurable for some time after supplementation ceased, leading to the conclusion that survival, multiplication and implantation of these most useful or organisms had occurred.

This experiment showed that levels of putrefaction and a decrease in the formation of carcinogenic materials was possible, simply by means of acidophilus supplementation, even in very elderly people.

2. Dr Barry Goldin and his colleagues wished to discover whether the differences which have been found in the relative microflora activity of vegetarians and meat eaters (i.e. fewer of the undesirable enzyme activities are found in the vegetarians) could be achieved by either altering the diet of meat eaters or supplementing their diet with lactobacilli.

They confirmed that a long-term commitment to a vegetarian pattern of eating, as is found amongst large population groups such as the Seventh Day Adventists, produced relatively low levels of activity from enzymes such as b-glucuronidase and nitro-reductase in the colon (Seventh Day Adventists also have a very low incidence of cancer).

When meat eaters (with relatively high levels of these enzymes) were placed on one- and two-month periods of increased fibre (bran, wheatgerm etc.) there was no effect on the levels of b-glucuronidase, nitro-reductase or azo-reductase. There was, however, a reduction in another undesirable enzyme 7-a-dehydroxilase.

When red meat was removed from the meat eaters' diet (white meat was still allowed) for one and two month periods again no difference was seen in these undesirable enzyme levels.

When, however, a normal omnivore eating pattern was allowed, including red meat, which was supplemented with *Lactobacillus acidophilus*, a marked reduction was seen in levels of b-glucuronidase and nitro-reductase. These improved levels returned to normal after the supplementation stopped.

In this study we see clear evidence of the potential benefits, in terms of a reduced risk of cancer, which can be offered by acidophilus. Dietary revolutions such as becoming a vegetarian clearly take far longer to achieve these particular benefits.

Remember that, in this study, when red meat was excluded, the eating of meats from poultry or fish were not stopped.

3. In another study, conducted by Dr G. V. Reddy in 1973, when volunteers previously consuming a fairly heroic level of over 100oz (6¼ lb/2¾ kilos) of meat weekly went on a diet, which completely eliminated all meat products for a month, there was a marked reduction of the presence in their GI tracts of the undesirable enzyme, b-glucuronidase.

In this study the dietetic changes were drastic, not just an elimination of red meat but a total ban on all meat products of all types. The results in terms of the change in behaviour of the bacteria of the region (specifically in relation to their production of enzymes) was, therefore, marked and was seen fairly rapidly.

Is it the meat or the fat?

4. In order to specifically identify the role of meat in the sort of changes described above, researchers led by Dr David Henteges, at the University of Missouri, asked ten volunteers to spend four months following a pattern as follows:

First month:	a 'control' diet (with 80g of beef once a day only).
Second month:	a meatless diet.
Third month:	a very high meat diet (300g of beef daily).
Fourth and final month:	'control' diet.

They concluded (and we do not agree with this conclusion) that the changes in diet made only a marginal impact on the *composition* of the intestinal fora, but acknowledge that this was probably because they made no other changes in the fat intake of these volunteers. The daily fat intake remained at around 80g during all the various dietary stages followed over the four month period (including the non-meat one).

We can get an idea of the changes which were observed to take place over the four months if we consider that levels of concentration of bifidobacteria remained, during all four stages, at a level of ten billion organisms per gram of faeces.

Bacteroides, however, showed some population changes, rising from an initial ten billion organisms per gram (when on control diet 1 and the meatless diet) up to a concentration of one hundred billion organisms per gram when the volunteers were on the high meat diet, remaining at this pretty high level for the final control-diet month.

Lactobacilli, in contrast, dropped from a level of ten million per gram in the first control and meatless months, to a density of only one million organisms per gram during the high meat diet, rising again to ten million for the final control month, when meat intake dropped.

Are such changes significant? The researchers suggest not. We would suggest that there is the potential for this to be of some significance. There was, after all, a major change in the overall balance resulting from a rise during the high-meat month, in numbers of bacteroides (from ten billion to 100 billion), as well as a fall in numbers of lactobacilli during that month (from ten million to a million). Another important thought rising out of this study is that the fat content of the diet is probably more critical than the level of protein consumed.

It also indicates once again that longer term studies are more likely to show changes of greater significance, a month is not long enough for a major impact. Finally it shows us that we should be paying attention to the *function and behaviour* of the flora rather than just their numbers (although this is clearly important) something which the researchers in this study acknowledge when they say: 'It appears that the metabolic activities of the bacteria, regardless of species, need to be examined under different dietary conditions in order to understand the role of bacteria in the aetiology [causation] of colon cancer.'

We should remember, when considering the evidence relating to meat eating and the ecology of the bowel, that much modern meat (especially beef) is very high in fat content, even when visible fat is trimmed. Since fat levels in the diet are proven modifiers of the behaviour of the bacterial flora, this should be worth considering. If meat is to form a major part of the diet then low fat versions should be sought, and these include game. Venison, for example, contains less than 4 per cent fat in terms of its body weight, whereas modern commercially produced beef can reach an astonishing 30 per cent of body weight as fat.

A choice is open to us in terms of the type of diet we choose, the risks we take and the protective actions we adopt. Acidophilus supplementation and reduction of fat (and perhaps meat) are shown to be most desirable examples of any preventive strategy we may adopt.

Tumour inhibition and the treatment of cancer with friendly bacteria

We have seen that the friendly bacteria can:

1. Enhance immune function.
2. Decrease the levels of potential carcinogens in the body.

3. Deactivate enzymes (produced by other bacteria) which have the ability to alter *possible* cancer-causing agents into *real* cancer-causing agents.

Now we will examine the evidence which shows that the friendly bacteria can also help in dealing with many forms of cancer and the side-effects of other forms of treatment, once cancer exists.

Amazing Bulgarian research by Dr Bogdanov

Perhaps the most startling and exciting information we can share with you relates to the many years of research conducted by Dr Ivan Bogdanov of Sofia, Bulgaria. He has produced conclusive evidence of the enormous benefits gained by treating existing carcinomas (of various types) with material extracted from specific strains of *Lactobacillus bulgaricus* (Strain LB-51).

It was, in fact, Dr Bogdanov who first isolated (in 1951) the antibiotic produced by *L. bulgaricus* (Strain LB-51) but he was not able to purify and stabilize this. It was some years before Dr Shahani found a way of doing this.

In 1956 Dr Bogdanov discovered that this same strain synthesized an anti-cancer substance which necrotizes (kills) tumour cells without being toxic to the person (or animal) being treated. These synthesized anti-cancer products cause no allergic responses as a rule, but actually seem to stimulate the immune system into greater efforts against the tumour. Solid tumours appear to respond most dramatically to this form of treatment.

It was in Bulgaria that the inital discovery was made of the potent anti-tumour activity of *Lactobacillus bulgaricus*. Dr Bogdanov and his colleagues isolated three different chemical fractions from *L. bulgaricus*, which were used effectively against sarcomas and ascites tumours.

The name 'blastolysin' was given to these extracts which were specifically found to be effective against tumours in mice. These effects were, however, not observed when tumour cells were exposed to the bulgaricus extracts in laboratory tests, and the conclusion was reached that: 'Blastolysin activates the animal's immunological mechanisms.'

Careful study by the Bulgarian research team showed that it was a component of the cell wall of *L. bulgaricus* (called a peptoglycan) which had the anti-tumour effect. This substance is present in the cell walls of many, if not all, lactobacilli.

Drs Friend and Shahani report that similar anti-tumour activity occurs when extracts of *Lactobacillus acidophilus, Lactobacillus casei* and *Lactobacillus helveticus* are used in treating sarcomas in mice. Dr Shahani found that it was *Lactobacillus acidophilus* (Strain DDS-1) which produced the strongest anti-tumour effect, again in mice.

The application by Dr Bogdanov of the substance he had discovered was originally performed intravenously, but more recently it has been found that there are advantages to its being taken by mouth. There appears to be rapid absorption and effectiveness when taken orally. The benefits do not diminish and, even after many months of use, it has been shown to produce no harmful effects in volunteers.

More importantly, perhaps, it was also found that this substance was well tolerated by patients who were in advanced stages of cancer, who had received radiation therapy and chemotherapy. In such individuals Dr Bogdanov comments on the 'stimulatory effect of the preparation on the regeneration processes in the organism'.

Proof of the effectiveness of these bacterial products is now overwhelming. The evidence we will quote relates to clinical trials which started in 1967 on human patients. The substance extracted from LB-51 synthesized products was called 'Anabol' and this actually contains the processed cell bodies of *L. bulgaricus* (LB-51). The amount of concentration of *L. bulgaricus* used in this way can be judged by the fact that an average daily dose of 'Anabol' contains the cell bodies and synthesized anti-tumour products found in between *40 and 60 kilograms* of Bulgarian yogurt!

Doses of between 10 and 15g of 'Anabol' are usually given daily in the first few days of treatment, with prolonged use showing a continuing mild anti-tumour effect. At last three months of treatment are suggested to achieve what Dr Bogdanov calls 'the anti-tumour effect'. In most cases (see below) the minimum length of time it was used was six months, with most patients continuing its use for between two and four years, the longest continual application being nine years.

When large doses, such as 30 to 40g are given, there have been indications of actual tumour disintegration. This can produce a toxic reaction as these breakdown products have to be cleared by the body's detoxification organs (liver etc.) which may already be overloaded and weak. For this reason, a modest, but regular and continuous, oral dose is preferred in severely ill patients.

The type of patients on which this method was tested can be divided into three groups:

A. The majority were severely ill people in what were considered to be terminal stages of their illnesses. All possible methods of treatment had been tried in these cases, with no beneficial effect. These had all been discharged from hospital and sent home wth purely symptomatic treatment being currently used (painkillers etc.).
B. Cancer patients who had had severe side effects from simultaneous chemotherapy and radiation therapy.
C. Patients whose bone-marrow had been damaged by previous radiation and chemotherapy, thus reducing their ability to manufacture white blood cells.

Among the 45 severely ill people (group A above) in a continuing study there were cases of pancreatic, thyroid, bladder, laryngeal, breast, stomach, lung, rectal, uterine, ovarian, brain and other cancers, including malignant melanoma, sarcoma, multiple melanoma and Hodgkin's disease. In the majority of those treated with oral 'Anabol' (as the only therapy) some had complete regression of their tumours and most showed a wide range of improvements, with no harmful side-effects.

The benefits continued even after six years of ongoing use. A typical case from this group is quoted from Dr Bogdanov's monograph published in Sofia in 1982 (see references).

S.K.B., male, aged 57.
Diagnosis: Multiple myeloma, cachexia, uraemia, coma.
Histological diagnosis: Plasmocytoma.
05.1968 — Diagnosis at admission: chronic nephritis. Intolerable bone pains and severe cachexia develop, with atrophy of muscle mass (the patient cannot lift his hand). X-ray shows multiple overlapping myeloma foci in bones. Diagnosis confirmed by needle biopsy. Gradually becoming somnolent, in the course of one week uraemic coma developed.

Results of laboratory analysis immediately before initiation of Anabol treatment: Hb44%, urea 160mg%, ESR 140/160m.

12.07.1968: Anabol treatment began as only therapy, patient

moribund, general condition improved rapidly in the course of one week. Patient recovered consciousness. Pains decreased.

In two weeks voice was restored. No spontaneous pains. Opiates discontinued. Urea 44mg%, ESR34/63mm.

30.10.1968: General condition improved constantly. Muscles of chest and arms gradually recuperated.

15.12.1968: Five months after initiation of Anabol treatment, general condition very good. The body muscles are almost completely recovered except for legs, which remind us of the recent severe cachectic state of the patient.

1.7.1969: In good general health. No clinical evidence of disease. Walks alone, using only a cane. Discharged from clinic. Lives in village. Takes care of himself, able to carry out work without strenuous physical exercise.

1.1.1970: Patient died of influenza.
Complete remission without evidence of recurrence continued for 17 months.

There are many more such cases quoted in Dr Bogdanov's booklet, all having been treated only with *L. bulgaricus* (LB-51) products.

Those cases described in B above showed marked improvements, some being able to continue with chemotherapy for their conditions. Those in group C, whose bone marrow had been damaged, showed rapid therapeutic effects, even in severe cases of very low white blood cell counts.

Dr Bogdanov summarizes his findings over many years of research and active use of this method by stating that: it has a *therapeutic effect* where there has been severe radiation and chemotherapy damage, and the benefits are seen within a few days. It has a *protective effect* against the harmful effects of other forms of cancer treatment, ensuring a maximum use of their potential for tumour destruction.

It has *a therapeutic effect* in incurable cancer patients in whom all other methods have failed and, therefore, Dr Bogdanov correctly asks, *why not use it before that stage is reached*? This is indeed a sensible question, deserving of a considered answer by those treating cancer.

Dr Shahani and Friend seem to feel the case is yet unproven since they state: 'Long-term controlled studies are required, however, to ascertain the benefits of dietary lactobacilli in preventing or reducing cancer in human subjects.'

We would suggest that in the light of Dr Bogdanov's experience and claims, and those of other researchers discussed below, there is no reason to delay use of this approach in treating cancer, above all since it is totally safe.

Other research

1. In research in Nebraska, Drs Friend, Farmer and Shahani prepared extracts from *Lactobacillus acidophilus, Lactobacillus bulgaricus* (several versions or strains) and *Lactobacillus casei* and used these extracts to slow the growth of tumours (Ehrlich's carcinoma and Sarcoma 180). When the mice in these studies were fed yogurt for seven days there was up to a 35 per cent reduction in the numbers of tumour cells.

Subsequent investigation showed that when yogurt cells were placed directly in contact with tumours there was a marked killing of tumour cells, with levels of reduction in these of up to 45 per cent being measured. These effects were no longer present after the yogurt cells had been heated showing that whatever the anti-tumour substances are, they are not able to withstand heat.

When different dilutions of yogurt were given to mice in their food the anti-tumour effect directly corresponded to the amount of yogurt consumed.

2. In further research into just what aspects of the yogurt cells could produce anti-tumour effects, Shahani and his colleagues separated different fractions (acidic, basic and neutral fractions) of the cells and tested these in laboratory culture dishes. No response was observed. In other words, in a laboratory dish, the yogurt cells seemed to have no potential against tumour cells.

However, when these same fractions were injected into mice with Ehrlich's ascites tumours, some showed a remarkable ability to destroy tumour cells. Thus it was found that when the cells themselves were injected (or their anionic fractions) there was a 38 per cent degree of tumour inhibition.

Why should these cells, or their fractions, be active in the body and not in a laboratory dish? Shahani and his colleagues suggest that: 'These anti-tumour principles may be 'inactive' when ingested or inoculated but later 'metabolized' or converted by the host into active components. It is also possible that these inactive components somehow trigger the immune system into action.'

This later suggestion (immune system involvement) is supported by researchers in Japan and Russia who have identified anti-tumour substances in the cell walls of *Lactobacillus bulgaricus*. Again, this was shown to be active when injected but inactive in a laboratory dish when exposed to tumour cells.

3. *Lactobacillus bulgaricus* was studied in Germany specifically to see whether anything in it, or produced by it, could explain the benefits it bestowed on animals with leukaemia. Mice with leukaemia which received injections of filtered freeze-dried extract of *Lactobacillus bulgaricus*, were compared with other mice with leukaemia who received no treatment, judged by the length of survival of these animals.

After mixed results initially due to problems with the cultivation of the bacteria and the filtration of the injectible product, consistently good results were achieved (described by the researchers as 'strongly positive') with a particular strain of *Lactobacillus bulgaricus*. This was strain CH-3 filant (available from Christian Hansen Laboratories, Copenhagen 2100) which is used in manufacturing what the researchers describe as a 'special ropy' type of yogurt.

It is not known whether this animal study has led to any attempt to use similar tactics in human leukaemia. It seems, therefore, that some strains of *Lactobacillus bulgaricus* contain or produce substances which have a potentially beneficial effect in leukaemia, and it is to be hoped that research will investigate this.

Chapter 11

The toxic bowel – toxic liver connection: dysbiosis in action

In the last chapter we saw something of the ways in which highly toxic substances can be manufactured in the gastro-intestinal tract from specific food residues, this process being engineered by an interaction between these substances and some of our normal bowel inhabitants. We also saw that a great deal of protection can be offered against such processes, and toxic chemicals, by various of the friendly bacteria.

We would now like to expand on this theme by explaining aspects of an all too common phenomenon, the toxic bowel, and some of its appalling health repercussions. In order to do this we need to firstly get rid of a number of unfortunate misconceptions:

1. There are many medical personnel who still believe that the mucous membrane of the bowel is not permeable, that whatever is undesirable in that region remains isolated from the bloodstream and eventually passes out of the body. This is a profoundly important misconception.

Evidence was established back in the early 1970s by Drs Warshaw, Walker and Isselbacher who showed unequivocally that what are called macromolecules can indeed pass through this barrier. This means that large molecules of partially digested food, as well as metabolic and bacterial waste substances are commonly able to find their way into our bloodstream where they may provoke a reaction on the part of the immune system.

2. Associated with this mistaken concept is another which maintains that when we eat protein foods the body invariably and efficiently breaks these down into their basic amino acid constituents, these then being absorbed into the bloodstream for

ultimate use by the body in a multitude of repair and building roles.

This is the ideal, but unfortunately it is not always met. The same Dr Warshaw and various of his research colleagues were able to show that large protein molecules, consisting of a number of still linked amino acids, called polypeptides, were able to reach the bloodstream from the GI tract and exercise their great ability to provoke antigenic responses.

3. This last misconception is compounded by the belief of many doctors who are aware that this is possible, but who yet maintain that it is at most a rare occurrence. In fact, it is now known to be a very common happening, and this suggests that toxins, bacterial proteins and breakdown products, as well as partially digested food particles may be entering the bloodstream with extremely nasty effects on the health of the body as a whole. Please remember that these often highly toxic particles place demands on the defence system of the body over and above those toxins which are normally absorbed from the GI tract, such as ammonia and phenols (of which more below) which need to pass to the liver for detoxification as a regular function of this organ.

If the bowel is in a toxic state, due to all or any of the many possible causes already touched on in previous chapters, the degree of self-poisoning (auto-toxaemia) may become a major feature of an individual's serious ill-health. Such a situation will always involve a microflora population which would be in a thoroughly unhealthy state.

Some repercussions of a toxic bowel

Before we examine the role of the liver in all this we might pause to add a few more unpleasant possibilities to the lengthening list of potential consequences of bowel dysbiosis:

1. The possibility exists for an overgrowth of *E. coli* in the bowel to provoke a *diabetic condition*. Unbelievable? The evidence is strong.

Dr LeRoith and his colleagues showed in brilliant research in 1981 that certain strains of *Escherichia coli* produce a substance which is indistinguishable from insulin. Our bodies produce insulin in order to control levels of sugar in the blood, excessive levels of which can cause untold damage. Many adult-onset diabetics (people with unnaturally high levels of blood-sugar) are shown to have apparently normal levels of insulin in their blood, but this just doesn't seem to be able to perform its task of clearing high levels of sugar.

Why should insulin not work? It is considered that a major cause of this may be that the insulin-like substance being produced by the *E. coli* strain may block special receptor sites on cells which insulin needs to reach in order to perform its sugar-controlling function adequately.

It is worth remembering that we have already seen evidence of the fact that a diet rich in fats and sugars will result in alterations in the microflora balance, leading to the likelihood of bifidobacteria decline and *E. coli* overgrowth. Repopulation of the bifidobacteria, by their direct supplementation as well as the use of bifidogenic foods and substances (see below) controls *E. coli* and helps to restore normal function.

It is interesting that a diet high in fat and sugars is recognized medically as being the worst pattern for people with a diabetic condition. It is also one of the worst patterns for the friendly bacteria.

2. Another bacterial inhabitant of the GI tract which can become excessively active if dysbiosis exists, and which has undesirable qualities, is *Yersinia enterocolitica*.

This has the ability to provoke the production of substances which attach to cells in the thyroid gland resulting in an *over-production of thyroid hormone*. Fully 80 per cent of patients with the serious thyroid condition, Grave's disease, have been shown to produce immune-system generated antibodies to yersinia.

Patients with high levels of yersinia are often found to develop conditions relating to auto-immunity (such as forms of arthritis and iritis) in which the body attacks itself instead of the invading organism, mistaking its own tissues for those of the invader. (See Chapter 6 on how this sort of mistaken identification by the immune system can lead to ankylosing spondylitis and osteoporosis.)

3. An immune system attack on undesirable bacterial colonies is also thought to be the cause of *ulcerative colitis* in which destruction of the mucous lining of the lower bowel is the end result.

4. Endotoxins are toxins which are produced by bacteria in the bowel. These are held by the bacteria which manufacture them almost like miniature time-bombs, since they are usually not released into our systems until the bacteria is destroyed (by some activity of the immune system or by friendly bacteria, for example).

One way in which undesirable bacteria can be deactivated involves specialized, highly complex, substances produced by a little-known part of the immune system called complement.

If the body is provoked into producing complement because of endotoxin-creating micro-organisms, the result would usually be the death of the bacteria, a beneficial event for us which would unfortunately be accompanied by the release of its toxins. These are capable of producing severe health problems.

Among the conditions so far identified as having just such a probable involvement of complement and endotoxins are *psoriasis*, other skin diseases such as *dermatomyositis, lupus erythematosus* and *pancreatitis*. In psoriasis, for example, an improvement is usually seen when an endotoxin binding drug, such as cholestryamine resin, which acts in the GI tract, is given.

Similar results have been seen when the herb sarsaparilla (*similas officinalis*) is given to people with psoriasis since this contains substances which bind endotoxins.

In one study described by Patrick Donovan ND, 92 patients with psoriasis were given sarsaparilla with a marked improvement in 62 per cent of these, and 18 per cent becoming clear of the disfiguring skin disease.

5. If partially digested proteins enter the bloodstream there is the probability of a strong immune response against these substances. Conditions as diverse as *migraine, eczema, learning disorders, behavioural problems, rheumatoid arthritis* and *thrombo-phlebitis* have all been linked with food protein sensitivity.

In hospital studies, all of these conditions have been shown to disappear or diminish when food antigens were controlled by the removal of offending protein sources (eggs, wheat and milk being the major culprits) or when a rotation type of diet was introduced in which these antigen food sources were eaten only infrequently. Rotating eating is described in Chapter 14.

We have already, in Chapter 8, seen the consequences of bowel dysbiosis in relation to yeast overgrowth, and the multiple forms of damage this can lead to. If we add to that some of the evidence discussed above we can see that the toxic bowel, dysbiosis, is a very real (possibly a major) cause of human suffering, and that attention to it by repopulation and restoration of the friendly flora is a supremely important contribution towards the recovery of health.

'Normal' toxins and liver overload

A wide range of non-carcinogenic, but nevertheless very toxic, substances are produced in the human digestive tract, ranging from ammonia to phenols, and the toxic amines discussed in the previous chapter. Many of these are automatically 'made safe' by the body's main organ of detoxification, the liver. The ultimate breakdown products, after detoxification by the liver, are recycled for use in the body, or are excreted in the faeces or urine.

However, when liver disease exists there is almost always some degree of impairment of this detoxification function, and unprocessed toxic products may then enter the circulation causing a range of unpleasant symptoms. Even the brain is made toxic by such liver failure.

In advanced cirrhosis of the liver, for example, anything from mild mental aberrations to coma may be experienced, as the central nervous system becomes loaded with toxic residues. The main chemical factor involved in liver-induced coma is excessive ammonia in the system.

Ammonia may be the end-result of a variety of processes which take place constantly in the body, including the changes engineered by bacteria (such as clostridia, eubacteria, peptostreptococci etc.) which alter urea in the body into ammonia and carbon dioxide.

Amino acids, the building blocks of protein, may also be acted upon by bacteria to form ammonia. This happens far more often when the amino acid compounds being broken down are derived from meat protein, than when they are derived from milk proteins.

When ammonia, formed in the digestive tract, is passed to the liver for detoxification it travels via what is known as the portal circulation. Reaching a healthy liver, the ammonia would be transformed into useful amino acids or back into urea, in both cases the ammonia would be detoxified. However, this cannot happen with any degree of efficiency when the liver is diseased.

Strategies

If the liver is diseased a number of medically therapeutic tactics may be adopted. These include limiting the quantities of protein eaten in order to help reduce ammonia build-up, a tactic which can, however, be counter-productive since it usually leads to further weight loss and weakness if prolonged.

Methods which cleanse the bowel may also be useful in order to remove putrifying residues from the intestines, for if these are left they will continue to create toxic substances which, after absorption, load the liver with excessive work at a time when it can patently not cope. Such bowel-cleansing may involve enemas, colonic irrigations or use of purgatives and laxatives.

Often antibiotics are also employed in such toxic bowel conditions in order to reduce any abnormal bacterial activity, although as we have seen any long-term use of these leads inevitably to further disturbance of the intestinal flora, probable diarrhoea and almost certain dysbiosis. We have available one other possible aid for the toxic bowel. The contribution to detoxification by the friendly bacteria when the liver is diseased (in the form of bifidobacteria and bifidogenic factors) and this approach can be extremely useful and is strongly recommended.

Treating the liver by feeding the bifidobacteria

Dr Rašić, the internationally acclaimed Yugoslavian authority on the subject we are examining, encourages the use of a bifidogenic diet. (Remember that foods or transient bacteria which encourage the health and performance of the bifidobacteria are called bifidogenic.) One medical bifidogenic material is called lactulose. Lactulose is a totally non-toxic substance formed from milk lactose (milk sugar) in an alkaline medium whilst being heated.

Although known since the 1930s, lactulose's effect in encouraging an increase in levels of bifidobacteria was not recognized until the mid-1950s, and it was not until 1966 that it was used by doctors when treating and controlling the liver condition *portal systemic encephalopathy*, where the brain becomes overloaded with toxins due to liver failure. When patients were treated in this way it led to a general improvement in their condition and a marked reduction in the levels of ammonia in their bloodstreams.

In many other studies it was found that lactulose had similar beneficial effects in cases of cirrhosis of the liver, with or without portal encephalopathy.

> Note: We are describing the known and researched effects of lactulose, although this is not generally available to the public, as an illustration of what a bifidogenic substance can achieve. Lactulose being unabsorbed by the body, simply passes through the GI tract and encourages bifidobacteria activity.

Since lactulose remains largely undigested as it passes through the small intestine it reaches the large intestine intact and there it becomes available for degradation by the friendly bacteria of that region as part of their feeding pattern. Here it is fermented into lactic and acetic acids by lactobacilli, bifidobacteria and other enterococci, encouraging an increase in these populations.

The increased levels of these bacteria and their acid secretions has the natural effect of reducing the pH of the large intestine, making it far more acidic. This increased acidity is the important key which actually prevents ammonia from being absorbed into the bloodstream, for when there is a high level of acidity (low pH) ammonia remains in its ionized form and it is not then able to be absorbed by diffusion into the bloodstream. Thus, the toxic load on the liver is reduced, helping to keep ammonia levels down in the system as a whole, including the brain.

A further benefit of the increased colon acidity is that organic acids so produced stimulate peristaltic action, encouraging more normal bowel movements.

The objective is more bifidobacteria

In general, Dr Rašić informs us, when lactulose is given it helps to re-establish a healthy bowel flora with an increase in bifidobacteria, lactobacilli and enterococci. At the same time a reduction takes place in the less desirable enterobacteria, clostridia and staphylococci. When such benefits are achieved, patients are usually allowed to eat additional protein, which of course helps to normalize their overall nutritional status.

In one study described by Dr Rašić the average level of lactobacilli (in 12 patients with cirrhosis of the liver) before supplementation with the bifidogenic lactulose formulation was 1 million per gram of bowel contents, while the potentially harmful *E. coli* were at a population level of 100 billion per gram.

After supplementation with lactulose, the levels of lactobacilli rose to between 10 and 100 million per gram while *E. coli* dropped to almost the same level as the lactobacilli. This is a far healthier ratio than existed before lactulose was provided and is dramatic evidence of what we should always try to achieve when there is bowel dysbiosis.

Lactulose is given in syrup form, diluted 50–50 by weight with the syrup. Initially 3 × 10g doses are given after meals, gradually building up to three doses of between 60 and 100g. The desired level of lactulose supplemented is that which achieves two soft bowel motions daily.

Don't go looking for lactulose

Before you rush out to buy lactulose, however, we should repeat again that this is not considered to be a food supplement by the FDA in the USA, and is not generally available except to medical practitioners. It is also very expensive.

We have explained its use in the research studies discussed above so that the potential benefit which it achieves, in cases of toxic bowel and subsequent liver overload, of improving the status of the previously depleted bifidobacteria can be appreciated.

There are other ways of doing this, of course, including the use of bifidogenic foods (Chapter 14) and supplementation with the friendly bacteria, including bifidobacteria themselves, and friendly transients such as *Lactobacillus bulgaricus*. And it is just these tactics which we advocate in such serious conditions of bowel and liver toxicity.

Treating the liver by supplying bifidobacteria

The increased formation of ammonia (as well as additional toxic substances) in the bloodstream of patients with liver disease is the result largely of bacterial action, and this is the cause of the potentially fatal coma frequently experienced by advanced cirrhosis patients. Standard medical treatment in such cases often involves the use of antibiotics such as neomycin, which (while effective) is both expensive and damages the intestinal flora.

By supplementing *Lactobacillus acidophilus* and bifidobacteria (as well as lactulose if available) a less costly, more physiologically sound and acceptable method of achieving similar ends is possible.

Proof

Drs Muting, Eschrich and Mayer conducted a trial using this approach involving 20 patients (8 with liver cirrhosis combined with severe portal hypertension and 12 with chronic hepatitis or cirrhosis). They found that by supplementing the patients with bifidus milk there was a decrease (usually within 7 to 10 days) of blood ammonia levels, serum phenol levels and other toxins (as well as reduction in excretion of free amino acids in the urine) to within normal levels, without any restriction being placed on the patients' protein intake.

Benefits from treatment of this sort are best if it is started in the very early stages of portal hypertension. In other liver conditions long-term benefits follow long-term use of this approach.

There is a clear advantage in such cases in using bifidus milk instead of just the lactulose approach, according to Dr Muting and his colleagues, which relates to their relative protein contents. Lactulose is a carbohydrate (derived from milk sugar) whereas bifidus milk contains around 30 to 40 per cent protein, a distinct advantage for the severely ill patient.

Dr Rašić also suggests that in order to assist an overloaded liver which is dysfunctioning, we should not only feed the friendly bacteria (with bifidogenic foods and bacteria, for example) but that we should also add new colonies of bifidobacteria via a supplementation programme. In one study, described by Drs Rašić and Kurmann, involving 33 patients with various degrees of cirrhosis of the liver, the patients were given a reconstituted bifidus milk product, containing viable bifidobacteria (Eugalen forte) for up to two years. Dosage was 10g three times daily, increasing to 3 doses of 100g daily.

There was a general improvement in the conditions of these seriously ill people (three of whom were children) a decrease in ammonia and free phenol levels, lowered pH (increased acidity) of the large intestine and a higher level of bifidobacteria.

Needless to say, cirrhosis of the liver is an extremely serious condition and the treatment programme should be monitored by a qualified health professional. It is to be hoped that such professionals will learn (if they do not already know) of the ability possessed by large colonies of bifidobacteria to flush phenols, amines and ammonia almost immediately. A strategy which includes massive supplementation of bifidobacteria as well as transient (*L. bulgaricus*, for example) bifidogenic bacteria and bifidogenic foods is suggested.

General detoxification

If there is no liver dysfunction, but there is a serious attempt at detoxification (fasting, drug detoxification, mono-diets, etc.) the use of additional supplementation with bifidobacteria would be expected to reduce the load placed on the organs of elimination and detoxification in much the same way as is seen in the far more serious cases of liver disease such as cirrhosis.

A diet which enhances the health and activities of the friendly bacteria should be followed (see Chapter 14 on the bifidogenic diet) as well as regular supplementation with *Lactobacillus acidophilus*, bifidobacteria and transient bowel inhabitants (which enhance the activities of the lactobacilli) such as *Lactobacillus bulgaricus*.

Together with this, some form of fermented milk product (containing live, actual cultures) such as live yogurt, kefir, acidophilus milk or cottage cheese, should be regularly consumed.

Chapter 12

How the friendly bacteria help in some other serious (and not so serious) health problems

We have in past chapters seen evidence that the healthy, abundant presence of friendly bacteria can help to:

- Enhance bowel health and function.
- Increase our vitamin status (especially vitamins B and K) and help us to digest a number of foods such as protein, fats and carbohydrates.
- Reduce chances of certain common forms of cancer in a number of ways.
- Reduce levels of potentially harmful cholesterol, cutting down the danger of cardiovascular disease.
- Enhance immune function (defensive capabilities) especially in the very young.
- Help to control (and often to destroy) potentially dangerous pathogenic bacteria, viruses and fungi in a variety of ways, in conditions related to vaginitis, the genito-urinary tract, mouth ulcers and overgrowth of organisms such as *Candida albicans*.
- Act as the ideal treatment in (and often be preventive against) cases of diarrhoea and food poisoning (of which there is more in this chapter).
- Keep levels of sex hormones balanced, with enormous repercussions of this on menstruation, fertility and osteoporosis.
- Reduce and detoxify poisonous burdens such as ammonia from the body when the liver is stressed.
- Reduce the chances of auto-immune disease such as rheumatoid arthritis and ankylosing spondylitis. . . .

And yet, there is even more.

Among the many other health problems which have been researched in relation to the benefits of the lactobacilli and bifidobacteria are acne, anxiety and depression, constipation, food poisoning, headaches and migraine, and protection against some forms of radiation.

It is also suggested that once the ecology of the mouth is disturbed, the friendly bacteria can be involved in tooth decay and gum disease. The evidence is as follows:

Acne and skin infections

1. Dr Robert Siver of Baltimore, suggests that, 'Lactobacilli are a safe simple, "80% effective", treatment for acne, especially in boys and girls under the age of 18'.

He points out that acne is the commonest skin disease of adolescents affecting as it does 30 per cent of girls and 44 per cent of boys. It accounts for almost one patient in five consulting a dermatologist. Acne is basically a sebaceous gland dysfunction, perhaps related (in any given case) to hormone imbalance, nutritional inadequacy (often zinc deficiency, for example), emotional stress and hygienic factors, in any combination or acting alone.

Dr Siver found, when treating gastro-intestinal conditions with lactobacilli, that very often there was a coincidental improvement in the patient's acne. His method was to supply two or three tablets containing *Lactobacillus acidophilus* and *bulgaricus*, three times daily with milk for eight days, followed by two weeks without supplementation, and then repeated if necessary. Skin care and hygiene was, of course, also encouraged.

The results described by Dr Siver were excellent, most cases showing benefits within the first two weeks, although some of the patients took as long as three months to respond.

In 300 cases treated in this way, Dr Siver achieved an 80 per cent success rate (half had 'reasonable' success, these patients were usually aged between 18 and 25; and the other half of the success group had what are described as 'excellent' results, and these patients were usually aged under 18). There was often a marked aggravation of the condition during the first few days of acidophilus supplementation.

2. Let us not forget the excellent results reported in Chapter 8 of successful treatment of herpes infection.

3. The Russians have developed an 'acidophilus paste' for the treatment of some skin problems. A Bulgarian report on this product informs us that this is used directly on inflamed areas, especially if pus is present. An older Bulgarian product, 'Biolactin', which was created from *L. bulgaricus* cultured on milk, was apparently used in a similar way.

The Russian product uses *L. acidophilus* instead, and is claimed to have antibiotic properties, and to be particularly effective against fungal infections, as well as *E. coli* bacteria.

Anxiety and depression

We do not suggest that anxiety and/or depression, both potentially extremely serious states, are always a result of factors which can be remedied by lactobacillus supplementation. However, there is evidence that in some instances this may be a useful strategy, as part of a comprehensive therapeutic effort, which is aiming at dealing with the causes of the problem.

Drs Friend and Shahani address the possibility of anxiety being alleviated when the lactobacilli are enhanced as follows:

'While there are no direct data relating lactobacilli to the alleviation of anxiety and depressive symptoms, indirect evidence suggests a possible relationship.'

They point out that one of the more recent developments in treating such conditions is the use of individual amino acids such as tryptophan (also phenylalanine).

'Since lactobacilli release (when they act on protein foods) essential amino acids during fermentation, fermented dairy products are rich in several essential amino acids, including tryptophan.'

Dr Hangee-Bauer suggests that as well as there being the strong possibility of deficiency (as mentioned by Shahani and Friend) in cases of anxiety/depression, there could also be toxic causes:

'Colonization of the intestines with lactobacilli reduces the amounts of endotoxins produced by certain bacteria, some of which have been linked with a wide variety of inflammatory and mental disorders.'

There would seem to be every reason to use this knowledge when faced with problems of anxiety and depression, since there is very clear evidence of many such cases having, in combination or alone, nutritional deficiency or imbalance, toxicity or allergy as a part of the cause, interacting with other, perhaps psycho-social, factors.

Anyone with depression, however, should always receive expert professional advice, whatever self-help measures are adopted.

Constipation

We have seen how the activities of the friendly bacteria enhance peristaltic activity. This is the muscular motive power which causes the contents of the intestines to move forwards, and which seems to be activated more efficiently when the acidity of the region increases, as it does when the lactobacilli are creating lactic, acetic and others acids as part of their normal activities of fermentation.

A study conducted by Drs Ferrer and Boyd successfully used yogurt as a means of improving the conditions of geriatric patients with chronic constipation. Their general health and skin tone were also observed to improve after this.

A Swedish study was carried out in which long-term hospitalized elderly patients with chronic constipation were treated with acidophilus milk, 36 women and 6 men aged between 50 and 95, all of whom had used various forms of laxative, were treated for varying periods by the addition to their diet of acidophilus milk.

The pattern under which the study was conducted was that the patients were divided into two groups, one of which received as much acidophilus milk as they wished during the breakfast meal, for seven weeks. The other group received no fermented foods at this time.

After a period during which neither group received fermented milk those who had initially received no acidophilus milk were allowed it for six weeks and those previously receiving it, were not.

A careful check was kept on the amounts consumed, the patients' bowel actions and the consistency of the faeces. There was a marked reduction in both use of laxatives and in the degree of constipation during periods when acidophilus milk was consumed (as determined by more frequent bowel motions of softer consistency). The best effect was achieved with an intake of 200 millilitres per day of acidophilus milk.

The added bonus of this method of treatment for these often nutritionally inadequate people was that they received, through the acidophilus milk, a source of bioavailable (predigested) nutrients. No adverse reactions were recorded.

Dental caries, plaque and gingivitis: are they 'caused' by the friendly bacteria?

René Dubos in his marvellous book *Man Adapting* (Yale University Press, 1976) sums up clearly the reasons for normal bacterial inhabitants of our body becoming harmful:

'Some form of pathological, or at least abnormal, state must exist before the indigenous biota (friendly bacteria) can multiply to such an extent that it causes detectable deleterious effects'.

In other words, if you create a situation in which such bacteria can flourish in a region of the body which they are not normally resident, say the mouth for example, the damage they cause can hardly be blamed on them. In evolutionary terms, humans consumed relatively small amounts of fermentable sugars until recently, and the rise in consumption of these over the past century or so has been matched by the rise in dental caries.

The causes of this disease seem to be multifactorial, requiring the interaction of several factors at the same time. These are:

1. 'Susceptible teeth', which term indicates that there exist pits and fissures which can entrap food particles, or which have low levels of naturally occurring fluoride on exposed surfaces, and where there is poor ability to control caries causing microflora.
2. An intake of dietary sugars which provides the major energy source for cariogenic bacteria, as well as the source of acids which demineralize dental enamel. The acid also provides the 'glue' (adhesive polysaccharide) portion of dental plaque, the 'cement'-like accretion found on teeth. Fermentable carbohydrate foods serve as this sugar source.
3. Bacteria which are acid forming and which adhere to the teeth in the plaque are the actual agents which produce the caries. *Streptococcus mutans* has been tentatively named as the major caries-producing bacteria, although other streptococcal strains have also been implicated.

Interest has also been expressed regarding the activity in the mouth of *Lactobacillus acidophilus*, the bifidobacteria and *Lactobacillus casei*, since these can all cause acid build-up in the mouth when suitably fed on a sugar rich diet.

Various organisms found near the gum margins, such as streptococcus and neisserei are related to gingivitis when oral hygiene is poor. Effective flossing and brushing techniques and a balanced diet which reduces sugar intake and allows the normal slightly alkaline saliva to do its job of keeping down excessive levels of lactobacilli, are all parts of the answer to dental diseases of this sort.

Are the friendly bacteria then the cause of these problems? Only if you blame the smell of an overflowing dustbin on the flies which gather round it.

The mouth is meant to be an alkaline area in which the ptyalin enzyme in the saliva can begin the task of digesting starches. If, through poor hygiene and a sugar-rich (and therefore acid) dietary pattern, we provide a happy environment for certain stray bacteria to flourish, we should not blame them.

Food poisoning (see also Chapter 3)

The incidence of food poisoning, frequently fatal, is increasing year by year in every country on earth, and it is not always associated with lowered standards of hygiene.

Modern fast food outlets often do not cook food for long enough to deactivate pathogenic micro-organisms.

- In August 1976, in the USA, 139 people became ill after a community picnic (approximately half of those attending). Investigation indicated that the cause was contaminated potato salad from which the 'causative' agent *Haemolytic streptococcus*, was isolated, as it was from the throats of 63 of the sick picnickers.
- In 1983 there occurred, over a period of a few months, in Maritime Canada, 7 adult and 34 infant cases of poisoning with *Listeria* (formerly bacterium) *monocytogens*, which causes the form of food poisoning known as listriosis. Investigation of the food source of this poisoning led to the discovery that infected coleslaw was the origin. The salad had been made with cabbages which had come from a farm where they had been fertilized with untreated sheep manure (the sheep were harbouring this bacteria) and stored in an environment suitable for listeria development.

- In 1983, 49 people were hospitalized in Massachusetts with septicaemia or meningitis caused by *Listeria monocytogenes*. Of these, 7 were newborn babies and the remainder adults; 14 patients died. The common source of infection was found to be a 2 per cent pasteurized milk. A second outbreak occurred soon afterwards in Connecticut, associated with the same milk source. Cows which provided this milk were found to have had mastitis infected with listeria, although no evidence of such infection could be found at the time of the investigation, nor could the normal dairy pasteurization procedures be faulted.

  There are currently a number of as yet unproven theories as to how these bacteria might be able to survive pasteurization. The authorities continue to look (so far in vain) for evidence of contamination of the milk in these outbreaks, after it has been processed. About 30 per cent of people infected by listeria die.
- In 1985 in Illinois 17,832 people were poisoned by *Salmonella typhimurium*; of these 14 died.

  The source of this poisoning was milk which had been contaminated through a mistake in the processing which allowed raw milk to mix with already pasteurized milk.

Is it just carelessness?

The cases mentioned above, including the tragic Illinois outbreak which was the worst mass food-poisoning in American history, might all be put down to unfortunate accidents, were it not for the fact that it now seems clear that standard pasteurization procedures are not capable of ridding milk of listeria contamination.

At least 4.2 and, in some tests, as much as 12 per cent of all raw milk in the USA is found to contain this micro-organism. The Massachusetts outbreak related to pasteurized milk despite the fact that other sources are blamed (these have never been specified).

Other outbreaks have been linked to pasteurized cheese in California, as well as pasteurized non-fat milk and cottage cheese in various experimental studies.

According to the Center for Disease Control, there are an

astonishing 40,000 reported cases of food poisoning in the USA every year of whom around 1,000 die.

This is only the tip of a very large iceberg according to the prestigious US News and World Report (18 April 1988) which says:

'Salmonella food poisoning is believed to afflict from 400,000 to 4 million Americans every year. It is rarely lethal but causes diarrhoea, cramps, fever and vomiting.'

Some researchers suggest that the figures of the CDC are very conservative and that upwards of 8,000 people die each year in the USA from food-borne infections.

Why is it happening?

Modern methods of animal farming (in most countries) promote the growth of salmonella and other potentially dangerous bacterial agents. Commercial feedstuffs often contain a variety of undesirable bacterial contamination (chicken food pellets contain a large percentage of 'processed' chicken faeces as well as powdered heads, feet and entrails of slaughtered chickens) thus repetitively reintroducing micro-organisms to the digestive tract of the animals feeding on such food. It is just such unnatural feeding patterns which have resulted in the current epidemic of *Bovine spongiform encephalopathy* which is affecting cows in Britain. This disease results in madness and complete loss of nervous system control in animals. It is derived from a virus-like agent which causes a sheep disease, scrapie, which was inadvertently transferred to cows when powdered sheep remains were fed to them in order to provide these gentle vegetarian creatures with a high protein diet in order to speed their growth and boost their milk supply. Many farm-raised animals are over-confined, fed, as described above, for rapid growth, dosed with antibiotics to keep levels of active disease down, frequently allowed to become smeared with their own and other animals' faeces in overcrowded transport vehicles, rapidly slaughtered in manners which often cause rupturing of the intestines and further contamination with faeces . . . all leading to the likelihood of the spread of bacteria into meats offered for sale in supermarkets.

Chickens, after slaughter, are plucked by rubber 'fingers' which beat the feathers off the carcass. These fingers, become smeared with excrement and this is literally beaten into the pores of the bird. The now filthy but defeathered chickens are thrown into vats where they literally float in a 'soup' of faeces, entrails, other carcasses and an astronomical degree of bacterial contamination.

The result is that at least 35 per cent of all factory-farmed chickens sold in British, European and American supermarkets, butchers' shops etc., are contaminated with salmonella. The level of contamination reaches 70 per cent in some cities.

In New York City, according to the Center for Science in the Public Interest's recent publication, *Guess What's coming for Dinner*(March 1987) the following levels of contamination with one or more pathogenic bacteria, all capable of causing food poisoning, were found after analysis:

Roast beef 52%, Chicken breast 53%, Ground beef 17%

The American Academy of Sciences estimates that 48 per cent of all salmonella poisoning in the USA (Britain would be no better) is caused by poultry contamination. They also estimate that half the human population is infected by either salmonella or campylobacteria every year.

Mention was made above of animals in poor conditions being regularly dosed with antibiotics, in order to keep them alive long enough to reach market in a reasonable state. This has led to many strains of bacteria, salmonella above most others, becoming resistant to antibiotics, a major and even potentially fatal problem for anyone who is poisoned by such an organism. The major strains of bacteria which can cause food poisoning are *Clostridium botulinum* (the deadliest by far) which affects the central nervous system, salmonella, staphylococci and of course listeria.

What about eggs?

The Center for Disease Control has found that a very large number of salmonella poisoning incidences relate to egg contamination with salmonella. The salmonella organism was transferred on chicken faeces on the shell in days gone by, but modern methods ensure clean shells. So where is the salmonella coming from? The CDC research indicates that they are *inside* the eggs.

Modern factory farming conditions are doubtless at the heart of this problem. If birds are infected internally (in the oviducts, where eggs are formed) the chance of contaminated eggs being produced is high. Chickens are also regularly dosed with antibiotics when kept in crowded, totally unnatural conditions, such as prevail in modern factory farms. Such medication makes the chances of highly resistant strains of salmonella developing even greater.

Should we, therefore, stop eating eggs? No says the CDC, just use the following tactics:

- Boil for 7 minutes, poaching for 5 minutes or frying on each side for three minutes kills any bacteria in the egg.
- Eggs which are only half cooked are seen to be potentially dangerous.
- Raw eggs are even more hazardous.

Not surprisingly egg producers discount the assertions of the CDC relating to egg contamination.

> Note: We urge you to remember the evidence in earlier chapters of the power of the antibiotics, such as those produced by *Lactobacillus bulgaricus*, which can deactivate salmonella. In these dangerous times, with food contamination a major problem as social patterns change and more food is consumed away from home, the use of proven safety measures such as regular supplementation with the friendly bacteria is clearly a wise strategy.

Tips for avoiding becoming a statistic

- Nearly 70 per cent of food poisoning occurs as a result of food eaten away from home.
- Regularly supplement yourself with friendly bacteria which, as we have seen in previous chapters, are capable of halting the activities of many of these pathogens (salmonella, botulinum) and of deactivating their toxic by-products which are the main cause of the symptoms of food poisoning.
- Buy 'organic' meat, from free-range animals, or game, rather than supermarket, factory-farmed meat.
- Cook food thoroughly and eat it as soon after cooking as possible. Never reheat cooked meat or fish.

- The centre of any meat being cooked has to reach 165°F/74°C in order to kill salmonella bacteria. Use a food thermometer to ensure this when cooking joints of meat.
- Never allow any cooked food to stand for longer than two hours at temperatures between 40°F/4.4°C and 140°F/60°C. This is the range between which bacteria breed and thrive.
- Wash fruits (especially strawberries) and salads (especially lettuce) thoroughly, and never eat such foods of which you are not sure, in terms of cleanliness.
- Use scrupulous hygienic methods when dealing with raw meats (barbecues for example). Wash and scrub hands, cutting boards, knives etc. and do not serve the cooked meat on plates which previously had raw meat on them.
- If a chicken has been on a work surface, clean the area thoroughly before any other food is placed in that area.
- Keep all raw food refrigerated, as this slows down any multiplication of bacteria such as staphylococcus.
- Frozen meats and fish should be thawed in a refrigerator.
- Raw fish (e.g. in Japanese restaurants) is dangerous unless absolutely scrupulous care is taken over hygiene. Do not eat raw meat or fish if possible.
- Never use eggs that are cracked.
- If tins containing food are in any way damaged, or show signs of being 'swollen' as though under pressure from gas inside, never, never eat the food in them. This is likely to be contaminated with botulinum which is the most dangerous of all food poisoning bacteria.
- If the contents of a can or jar of food are obviously mouldy, or there seems to be any foam-like quality to the contents, or the smell is at all suspicious, get rid of it immediately and protect animals from any chance sampling of the contents.

Headaches and migraine

These conditions are often associated with depression and anxiety, and also with a variety of food and digestion related problems.

Monica Bryant B.Sc., of the University of Sussex, declares that the use of probiotics shows, 'consistently good results with headaches and migraines which are diet related'. One of the normal inhabitants of the human bowel (though in small quantities) is *Streptococcus faecalis*. This has the ability to alter an amino acid (amino acids are elements in many forms of protein in our food) tyrosine into a substance, tyramine, which is known to cause migraine headaches. Tyramine is found in some cheeses, chocolate and many wines in large quantities, all these are well known for their ability to trigger headaches in sensitive individuals. If there were an imbalance in the populations of indigenous microflora, with faecalis being more in evidence than usual, this sort of increase in tyramine's presence would be more likely than ever to occur.*

Since some forms of headache are diet related, and even more have roots in stress and anxiety, the use of probiotic products (which do not contain *Streptococcus faecalis*) which can restore a healthy balance to the intestinal flora, would seem to be both safe, and probably useful, for people so afflicted.

Radiation protection

Research on guinea pigs which were deliberately irradiated, is reported by Dr Shahani. Some were also given acidophilus milk. Those receiving no acidophilus milk in their diet produced a great number of severely deformed offspring. Those in the experimental group, who did receive the milk, not only had no deformed offspring but produced litters in which the young developed better and gained more weight than those from the non-acidophilus group.

It is probable that the enhanced general nutritional status (and reduction in toxic elements), which we now know the friendly bacteria to be capable of ensuring, is the reason for the benefits seen in these unfortunate animals. Are there any benefits for humans? Moira Timms and Zacharia Zar believe so, and explain, in their book *Natural Sources* (Celestial Arts, California 1978) that the consumption of between a pint and a quart of yogurt, or soured milk, daily will help dramatically to remove residues of radioactive strontium 90 from the bowel.

*We should mention that *S. faecalis* is listed in the 'bible' of bacteriology (Bergy's Manual [Volume 2]) as also being an agent in urinary tract infections and subacute endocarditis (inflammation of the heart).

Conclusion

We can now add to the lengthy list of conditions which are helped by the friendly bacteria, with which we started this chapter. We can include such diverse problems as acne, constipation and, possibly, depression and anxiety-related conditions, including some forms of headache. Protection against food poisoning and radiation is a further bonus. Among other speculative suggestions (as yet unproven by studies) of health benefits are some which link improvements in gout, some forms of arthritis and rheumatism, and duodenal ulceration to the use of these marvellous organisms.

Chapter 13

Infants and the friendly bacteria

There was a time when breast-milk was undoubtedly best for infants. Now, however, even that splendid food source seems to have become contaminated by the damage we have done to the ecology of our planet.

There is strong evidence that the very special qualities of breast milk have declined, worldwide, probably due to contamination by man-made pollutants, with the result that the intestinal flora of infants today differs in its make-up from that found in infants only 30 years ago. This has some potentially serious consequences.

We have seen in Chapter 2 that the predominant bifidobacteria strain found in the colons of breast-fed infants used to be *Bifidobacterium infantis*. More recent investigations now show a dramatic shift in this picture, with other strains such as *Bifidobacterium bifidum* and *B. longum* now predominating in the colonies of breast-fed babies.

Does this change in predominant strains matter?

A group of French doctors — Beerens, Romond and Neut — have studied these changes and have concluded that the colonizers of the bowels of infants fed on breast milk differ immensely from those fed on cow's milk or formula feeds. The most obvious visible (and odoriferous) differences are pointed out by these researchers: 'Breast feeding induced liquid faeces with a 'cheesy' odour and acid pH (around 5.0). The faeces of artificially fed infants look like those of adults in consistency, odour and pH (between 6.0 and 7.0).''

They tell us that the numbers of bifidobacteria in groups of differently fed infants may often appear to be similar, but the strains of bifidobacteria differ, and this carries with it serious health implications.

'The main difference between infants fed cow's milk and breast-fed infants is the presence of bacteroides, clostridium and *E. coli* in high numbers in infants fed cow's milk, which do not decrease with age. The breast-fed infant possesses a natural defence (through particular strains of bifidobacteria) against *E. coli*, bacteroides and clostridium which grants the infant resistance to gastroenteritis.'

After examining a variety of possible food sources which might enhance bifidobacteria colonization of the GI tract, the French doctors concluded that: 'None of the other mammalian milks favoured the growth of bifidobacteria strains (*B. infantis, B. bifidum* and *B. longum*). This suggests the presence of a specific factor which we have termed 'BB factor'. This substance in human milk, which stimulates bifidobacteria growth, is not identified, but the French team claim it to be heat-stable and not a protein.

They then ask a pertinent question. Do the differences which we see in levels and type of bifidobacteria simply reflect differences in the health of these babies, or are they actually contributing towards these health differences?

Evidence

They suggest that health benefits which derive from a healthy flora are clearly demonstrable, and given the example of the very first milk a baby receives from its mother. This first milk (the colostrum) contains important substances which fight hostile bacteria (*E. Coli*, bacteroides and clostridium for example). This encourages the development of bifidobacteria colonies, which in turn increases the acidity of the colonized region.

While manufacturers of infant formulae try to mimic aspects of these qualities, such formulations are based largely on cow's milk and this simply cannot sustain those bifidobacteria colonies in the baby's bowel.

So we are left with evidence which points to particular strains of bifidobactria contributing to the health of the infant rather than the presence of bifidobacteria merely being a reflection of the child's health. The change in type of bifidobacteria and the other differences in bacterial populations which have been observed in infants fed on different foods undoubtedly contributes a good deal to the variations in health which are seen when we compare the relative health records of breast-fed and bottle-fed babies.

While this observation is demonstrably true, it may be less the case nowadays for we are aware that bacterial colonization patterns have changed over the past few years, even in breast-fed babies.

Dr Rašić's opinions

The leading researcher in the world of bifidobacteria is Dr Jeremija Rašić, and he believes that environmental factors have been the cause of this change:

'The possible explanations for the above changes may be . . . a consequence of increasing contamination of the human environment including air, water and foods (petrol, radioactive substances, antibiotics, pesticides, toxic minerals etc.).'

He sees that the natural environment has changed and that this has had direct consequences on the ecology of the bacteria of intestinal flora, especially the bifidobacteria, since they are far more sensitive to external and internal influences than are other bacterial groups. Dr Rašić described a study conducted on the composition of intestinal flora in infants for nearly 30 years (German study).

The continuing trend observed is of a decline in numbers of, and changes in, strains of bifidobacteria in breast-fed babies accompanied by a rise in levels of undesirable pathogenic organisms. He explains that between 1959 and 1966 bifidobacteria were cultured from the stools of all the breast-fed babies who were examined. During the period 1969 to 1972 as many as 5.8 per cent of babies examined had no bifidobacteria in their stools, and this level rose to 10.3 per cent in the group of infants examined between 1974 and 1977.

Similar trends were observed in premature babies studied over this time scale. Premature infants have a naturally lower level of bifidobacteria than full-term babies anyway, but the trend towards even lower levels was also demonstrable in this more fragile group of new-born infants, over the years of his study.

In contrast to the decline, year by year, in friendly bacteria in infants' GI tracts, Dr Rašić has found steadily increasing levels of klebsiella and *E. coli* (see Chapter 11 for dangers of *E. coli* overgrowth), both potentially disease-causing, in these same groups of babies.

Dr Rašić reports a worrying additional change: these strains of disease-causing bacteria are now commonly found to be resistant to antibiotics. A further change which he has observed is a gradual rise in average levels of pH of the GI tract of babies, indicating lower levels of acidity in the large intestine and consequent dangers of undesirable bacterial and fungal overgrowth.

Other researchers in Britain, Germany and France, have confirmed Dr Rašić's findings, and have extrapolated this trend forward in time, predicting that a steady erosion of both numbers and strains of the normal populations of the bowel will occur, bringing with it a tendency for increased levels in the bowels of infants (even breast-fed ones) of disease-causing micro-organisms, with all the consequences this can have in terms of diseases such as gastroenteritis.

Something has happened to breast milk, but what?

Dr Rašić believes that *B. infantis* is a less 'robust' bacteria than some other bifido strains (such as bifidum and longum) saying that they (infantis) are very slim rod-like organisms, which are very difficult to cultivate. Thus, the more sensitive infantis strain seems to be less able to cope with environmental pollution, despite its superb ability to deal with invading micro-organisms such as salmonella.

One study which has enormously worrying implications was conducted by Drs Arnold Schecter and Thomas Gasiewicz, who examined the relative levels of toxic chemicals such as polychlorinated dibenzo-P-dioxin (PCDD) and polychlorinated dibenzofurans (PCDF) in human breast milk in such diverse geographical locations as Vietnam, Canada and the USA. Their findings are cause for concern, for they show that levels of contamination of human milk with these dangerous chlorinated chemicals has occurred to a wide extent, and that infants are now in the front line of receipt of these toxins.

Levels of toxic breast milk in the south of Vietnam were far higher than safety levels dictate, as a result, it is believed, of herbicides sprayed on that unfortunate country during the war between North Vietnam and the USA. Levels of these potential carcinogenic toxins found in US mothers' milk was found to be at a level which is lower than in Vietnam although, 'nursing infants may be consuming levels which are near or above recommended safety values'.

The levels of body fat toxicity found to be present in the mothers tested were considered to be equal to that which is sufficient to produce the disfiguring condition of chloracne in industrial workers who become exposed to similar chemicals.

A report on this contamination of breast milk in *USA Today* (18 December 1987) quotes Dr Philip Anderson of the University of California, an authority of drug-tainted breast-milk, as saying: 'The longer they breast-feed, the more the baby gets. We cannot permit the contamination of our most precious resource to continue.'

Dioxins and furans do not only derive from the spraying of Agent Orange-type herbicides, such as took place during hostilities in Vietnam. There are other more mundane sources, such as garden pesticides, wood preservatives, chemicals used in paper production and in some household paper products as well as emissions from garbage incinerators. The chemicals have been found in atmospheric pollution, tainted water in some urban water supplies, as well as seafood fished from tainted seas, rivers and lakes.

What does this breast milk contamination mean in health terms?

Dr Schecter and colleagues urge long-term observation of babies fed on contaminated breast-milk, because the chemicals mentioned are specifically worrying owing to their ability to cause cancer in experimental animals. Such long-term observation would monitor the development of serious illness as the years pass, and would doubtless try to link any such changes with the ingestion, at the breast, of toxic substances.

Dr Rašić, however, as we have noted, has already looked at the effects on bowel ecology of more subtle toxins than the destructive herbicides used in the Vietnam war. He has found a decline in those bacteria (specifically of *Bifidobacterium infantis*) which can prevent infant diarrhoea and an increase in those bacteria which can cause it.

One of the major causes of death in babies is uncontrolled diarrhoea. This is especially likely when the infant is already malnourished or has been subject to recurrent vitality-sapping episodes of ill-health. This is most evident in third world and underdeveloped countries but is still a major cause of death in infants in the principal industrialized countries. Close to a thousand children under the age of one year die each year in the US from the effects of diarrhoea.

The bacteria most involved in this sort of illness are enteropathic *E. coli* which attacks either the small or large intestine, where it does its damage via production of toxins. Similar regions are invaded by various strains of salmonella which attach themselves to the mucous membrane and try to penetrate it. Once this is achieved it creates its damage by direct spread. Shigella does its work in the large intestine where it can produce symptoms through direct invasion or toxin production.

These and other pathogens are, however, quite capable of being controlled by a healthy flora when this includes bifidobacteria strains such as infantis. Tragically this particular strain, as Dr Rašić has shown, is in decline in modern breast-fed infants, probably due to toxic contamination of the mother.

He says: 'The sensitivity of *B. infantis* compared with other species does not invalidate the value of their freeze-dried preparations, containing a large number of viable cells, for babies. The regular consumption of a large number of *B. infantis* (100 million to 1 billion per day) ensures a steady supply of these friendly bacteria to the intestinal tract and by their 'mass effect', and some metabolic activity, these organisms help indigenous organisms to combat harmful bacteria, including pathogens.'

Of course, other strains of friendly bacteria, such as *L. acidophilus* and *L. bulgaricus* can also inhibit and deactive these pathogens by means of their antibiotic production, control of acidity levels etc.

If we are concerned to make up for the decline in bowel ecology which has taken place in recent years, and thereby to reduce the dangers of bowel infections, the course suggested by Dr Rašić, of regular supplementation with bifido bacteria, is the one we urge for all infants in this increasingly dangerous world.

CAUTION

L. bulgaricus and *L. acidophilus* should not be given supplementally to infants unless such a course of action is recommended by a well-informed physician, since some strains of these micro-organisms can produce forms of lactic acid ('D' form instead of 'L' form) which are poorly tolerated by infants. Supplementation with *B. infantis*, is however, safe for infants.

Other health benefits for infants from friendly bacteria

- Dr Rašić has found that when underweight infants, especially those who have been artificially fed (not breast-fed, in other words) are supplemented with bifidobacteria, there is an increase in nitrogen retention which results in a tendency towards achieving normal weight gain.

  Similar benefits are found when *Lactobacillus acidophilus* is supplemented to artificially-fed infants.

  This does not mean that normal weight individuals would gain weight if such supplementation were introduced, rather it indicates that when nutrients are not being suitably absorbed this factor can be normalized by repopulation of the GI tract with friendly cultures.
- We should also recall (Chapter 10) that the bifidobacteria have the ability to stop other bacteria from altering nitrates into nitrites, a process which could result in iron and oxygen depletion for the infant. The deactivation of carcinogenic chemicals by friendly bacteria applies just as much to the infant bowel as to that of the adult. Calcium absorption is enhanced by the presence of friendly acid-producing bacteria, a most important consideration for the growing child. Similar benefits are seen as far as digestion of milk sugars and proteins are concerned.
- B-vitamin production is a further benefit for the infant which a healthy presence of the bifidobacteria deliver.

A good start in life: Infant supplementation with friendly bacteria

Because of the need which infants, even breast-fed ones, have for a sound bowel ecology, some manufacturers are now marketing special formulations of friendly bacteria with which babies can be safely supplemented. We cannot stress too much just how important we feel such a strategy to be. Ideally these should contain the strain *Bifidobacterium infantis*, the predominantly friendly micro-organism of the healthy breast-fed baby. All bottle-fed babies should be so supplemented and we now urge breast-fed babies to receive similar supplementation, doubly so if the nursing mother is in anything but the very best of health, based on Dr Rašić's findings of a decline in *B. infantis* in modern children.

Other early supplementation which can assist in repopulation of the infant (and adult) bowel with desirable friendly bacteria could include another strain of bifidobacteria, this time *B. bifidum* (strain 'Malyoth'). We will list directions and quantities in Chapter 17.

Streptococcus faecium and infant food

British research by Drs Bullen and Tearle, points to the ways in which another common inhabitant of the GI tract appears to help bifidobacteria activity. This is *Streptococcus faecium*, which Dr Bullen has found to contribute enormous quantities of acetic acid, a definite bonus for bifidobacteria.

Some strains of this common (if minor) inhabitant of our bowels, have the ability to cause food poisoning. However, most are safe, according to European research, which is just as well since some manufacturers are now including large amounts of *S. faecium* in their supplemental lactobacillus mixtures, mainly because these cultures are cheaper and easier to cultivate than the friendly lactobacilli.

Such supplements are found on sale in health food stores and pharmacies. We will be examining this trend of using strains and cultures of questionable value, and the many implications this has relating to health, in Chapters 15 and 16. It is there that we will consider possible problems rising out of the use of modern commercial methods for the manufacturing and marketing of cultured foods, as well as those cultures (such as *S. faecium*) which are now being promoted as having health-enhancing potentials.

Research in India by Dr Batish and colleagues investigated the toxic potential of an undesirable strain of *Streptococcus faecium*, SF-100, which was found in dried infant food samples. They were able to show that in conditions of slight alkalinity and at temperatures of 37°C/98.6°F the enterotoxins produced by *S. faecium* were at their maximum production after eight hours.

The implications for human babies is not clear, except that should this strain find its way into an infant's digestive tract in any quantity, and should the normal flora be unable to deactivate these bacteria or detoxify these toxins, for any reason, unpleasant complications could ensue. We question the wisdom of using *S. faecium* in supplementation when the reasons for its use are often based on purely commercial considerations.

Recent research in Sweden (*The Lancet*, 18 March 1989, 'The Prepuce: A Mistake of Nature?', Winberg, J. et al.) suggests that the vogue for circumcision which is known to prevent 90 per cent of urinary tract infections in boys could be better achieved by what they describe as 'manipulation of the bowel flora'. The Swedish believe that by encouraging exposure of infants to benign (friendly) bacteria, a more physiological approach to removal of the prepuce would be achieved.

> We believe that *S. faecium* supplementation should only be undertaken (especially where infants are involved) under the guidance of an experienced physician.

Chapter 14

Enhancing the bowel flora through diet, supplementation and stress reduction

By now we should be aware of the range of disasters which an unbalanced bowel flora, and bowel toxaemia, almost certainly leads to. We need, if disease prevention and health promotion are priorities, to adopt strategies which both reduce the toxicity of the bowel and enhance the health and well-being of the friendly bacterial colonies which offer us so much in the way of protection.

There are positive and negative influences to consider when it comes to trying to enhance the status of the bowel flora. We need to stop doing those things which are disturbing the friendly bacteria, and we need to concentrate some attention on providing them (and us) with the right sort of nutrients in order to achieve the healthy internal ecology needed for good health.

We have touched on some of the major negative influences on the bowel flora and these include toxic food substances and materials including pesticides and herbicides, excessive fat and refined sugar in the diet, as well as drugs such as steroids and antibiotics, which destroy the ecological balance between the various friendly cultures and ourselves.

Anything which encourages particular bacterial cultures to become dominant at the expense of the normal flora is something which should be avoided. Evidence shows that a variety of foods and food types in the diet are preferable to a strict unvarying pattern of eating, at least as far as production of a balanced flora is concerned.

Drew Collins ND, writing in *The Textbook of Natural Medicine* puts it this way: 'Contrary to popular belief, frequent alternations of the chemical nature of the intestinal contents makes bacterial populations more responsive . . . as the diet is changed from pure protein to pure carbohydrate sources [of food] and back again, the flora exhibit an increasing tendency to become uniform, crowding out ubiquitous bacteria.'

Thus, alternation of dietary content brings about the prevention of any one type of bacteria dominating the others. When this concept of changes in the diet from high carbohydrate to protein and back again, was tried out on monkeys and kittens, a marked improvement in spirits and behaviour was observed.

So we need a varied diet for general colon health. We also need to include foods which are bifidogenic, which will encourage that most useful group to flourish, and this type of diet will be discussed below. Before doing so, however, we need to think about the effects on the ecology of the bowel of emotional stress.

Stress

Emotional stress has a remarkable effect on the bowel. Local blood vessels (arterioles) can become cramped and actually go into spasm as a result of anxiety or other stresses involving the mind.

When we are aroused emotionally, or suddenly alarmed by an event which calls for swift action, we enter the first phase of what is known as the general adaptation syndrome (GAS). This 'alarm' stage of the GAS is characterized by an instant release into the bloodstream of the hormone adrenalin (now known as epinephrine) which has a multitude of almost instantaneous effects. Among these are a release from the liver of stored sugar, to fuel the many responses of the body to whatever demands have been placed on it; a tensing of voluntary muscles in preparation for action ('fight or flight'); a speeding of heart rate and increase in blood pressure to assist circulatory supply for whatever musclular response is forthcoming; sweating starts, the mouth becomes dry, pupils dilate and, of interest to us . . . digestive processes cease, or slow down dramatically, as blood is diverted from this region to the muscles.

Should such an 'alarm situation' allow suitable action, which is then forthcoming (running away from, or successfully defending ourselves against, physical attack, for example) then the preparations the body had made (as listed above) would have been used appropriately. Muscles will have used the newly released sugar supplied by the enhanced circulatory activity and no ill-effects would ensue. All the listed changes then normalize as the body reverts to unstressed function.

If, however, no such 'fight or flight' response is possible, if the 'alarm' call of the stress event allows for no appropriate action, if for example we are in an unpleasant confrontation with an employer, or are faced by irritating and stubborn bureaucracy, or are stuck in a train, plane or car with no obvious end in sight to an uncomfortable delay, or if we create repeated anxiety through our own self-generated fears, hates, jealousies etc. (and we are sure you can provide other more immediately appropriate examples) then the arousal (adrenalin release and all the consequences) which is created is usually not suitably expressed, or used.

If such a pattern is often repeated, as is common in modern life, a change takes place in the alarm reaction. Instead of there being a repetitive recurrence of the pattern described above we enter phase two of the GAS. This is the adaptation stage (so called because it is during this period that the body adapts to the repetitive demands made on it) in which the responses become chronic rather than acute.

By this time (perhaps months or years into a pattern of repetitive stress) the body responses could include permanently tense muscles, headaches, high or low blood sugar levels (diabetes or hypoglycaemia), high blood pressure and tendency to palpitations, indigestion and/or stomach ulcers and almost certainly a degree of bowel dysbiosis due to the constant influence of negative circulatory and secretory changes brought about by this region's adaptation to stress.

This pattern of physical alteration is most obvious in the colon, for a number of complex anatomical reasons, and the result is of actual damage to the inner surface lining of the digestive tract, which can lead to local swelling, inflammation and at times to localized minute haemorrhages.

Such changes, not surprisingly, have a variety of effects on the health of the local flora. Depending upon which bacterial types are predominant in such a damaged area, and of course on the dietary contents of the region at the time, a variety of toxic chemical substances can be formed, some of which could pass through the now damaged mucosal lining into the blood stream with allergic and inflammatory consequences. There would probably also be an immune system response to the various stages of such inflammation. Stress, therefore, needs to be removed from the scene, or at least modified, as far as is possible if we are to restore, or more importantly, to retain, a healthy flora, and methods such as breathing, relaxation, meditation and visualization techniques are all useful in helping towards this end.

Anyone with chronic bowel dysbiosis should examine which areas of their lives can be modified to reduce stress, and should, along with practical measures, such as supplementation with friendly bacteria, consider the introduction of stress-reducing methods such as relaxation exercises. Just as it has now been shown, unequivocally, that the mind is capable of directly influencing, and indeed controlling, almost all aspects of the immune system, so its influence can be seen to extend, albeit indirectly but nonetheless extensively, to the inhabitants of our GI tracts.

Enhancing the microflora with other micro-organisms

As we saw in earlier chapters it is possible to enhance the health and function of colonies of resident bacterial flora such as the bifidobacteria by the simple means of regularly allowing transient organisms, such as *L. bulgaricus* and *S. thermophilus* to pass through the tract.

We suggest that in any attempt to enhance, restore or protect, the indigenous friendly bacteria, such bifidogenic bacteria should be regularly consumed. Foods which contain them, such as yogurt, can therefore act as a source of nourishment, or can induce beneficial environmental modification, for the bifidobacteria and *L. acidophilus*, as well as providing nutrients for us. If, however, a sound food source of some of these bifidogenic bacteria (such as live yogurt or other cultured milk products) is not available, then we recommend regular direct supplementing with cultures such as *L. bulgaricus*.

Food influences on bifidobacteria

A diet which is rich in complex carbohydrates such as fresh vegetables, pulses (the entire bean family), seeds, nuts and whole grains (wheat, oats, rice, millet etc.) and which includes a reasonable amount of fermented milk (or soy) products, provides the best basis for a healthy intestinal flora. This suggests that either a macrobiotic or lacto-vegetarian diet is capable of providing the most desirable nutritional background for bowel health, and by extension, of general health.

The evidence of health authorities world-wide supports this suggestion; for people who follow such patterns of eating enjoy superior health and much lower levels of chronic disease, including a lesser incidence of cancer, high blood pressure, cardiovascular disease and arthritic changes.

If animal protein is included in the diet then at least low-fat sources should be sought, with fish, poultry (minus skin which is fat-rich) or game as the safest sources of meat.

Specific bifidogenic foods and factors

Research by Dr Rašić and others has shown that some foods and food constituents have a specific ability to enhance bifidobacteria activity. First and foremost, modified forms of lactose (milk sugar) seem to have a marked effect on the health of the friendly bacteria. This confirms the suggestion we have made that *real* acidophilus milk, kefir and natural live yogurts are highly desirable as regular foods for anyone concerned with such microflora enhancement.

Fermented soy products such as miso and tofu, so much a part of macrobiotic eating, have similar benefits to offer.

Cow's milk, as such, contains imbalances (as compared with human milk) which can actually inhibit some of the friendly bacteria. The high protein levels and excessive calcium phosphate are two examples of this. Cow's milk contains far more protein than human milk, which is only to be expected since its purpose is to enhance the dramatically quick growth of baby cows, not the relatively slow development of human infants.

However, if cow's milk is modified in certain ways it can then become bifidogenic. One such modification, apart from the fermentation processes induced by friendly bacteria, which improves the bifidogenic quality of cow's milk, involves replacing the fat content (in skimmed milk, for example) with vegetable sources such as linoleic and linolenic acid.

Heating milk is another way of creating bifidogenic alterations. Lactulose, which is a combination of fructose (fruit sugar) and galactose, is created by heating milk, and it has been shown to dramatically increase levels of bifidobacteria in the stools of artificially fed (not breast-fed in other words) babies. The amount of lactulose given daily should be between 1g and 1½ g per kilo of body weight, so that it makes up between 1 per cent and 2 per cent of the baby's diet. Within one to three days marked growth of bifidobacteria is observable with this supplementation.

Dr Rašić tells us that the human digestive system can barely digest any of the lactulose, which passes almost intact to the colon where it is happily devoured by bifidobacteria, *L. acidophilus* and other organisms, reducing it to highly desirable lactic and acetic acids.

In adults who have an alkaline condition in the bowel (undesirable in the extreme) supplementation with lactulose promotes friendly bacterial growth and subsequent acidification of the region.

Lactulose supplementation in cases of liver cirrhosis is now a standard medical treatment, since it allows the liver detoxifying powers of the bifidobacteria to be enhanced. This approach is, however, one which physicians only can introduce. We can ensure a healthy bifidobacteria presence by simpler means (below).

Bifido factors

A substance which has the ability to promote bifidobacteria activity has been isolated in relatively large amounts from human milk (as well as from human secretions such as saliva, semen, tears etc. and in small quantities from ewe's, goat's and cow's milk as well) and is termed bifidus factor 1. Its name is the daunting, *N-acetylglucosamine-containing saccharides*. A second bifidus factor is found in human milk, cow's milk, pancreatic extract, insulin. It is also available as the result of enzymatic action on proteins such as cheese, when enzymes such as pepsin, papain and trypsin can get to work on such proteins.

The bifidus factors have been successfully used to enhance bifidobacteria activity in infants and adults.

Some plant foods have also been found to have specific beneficial effects on bifidobacteria status, including carrot, potato and maize extracts.

Dr Rašić's views: Dr Jeremija Rašić suggests that we should regularly consume cultured milks, containing bifido and acidophilus bacteria, thus ensuring a steady supply of 'starter' bacteria to the intestinal tract, helping to maintain a proper balance of resident bacteria.

He points to the enormous difference found in bacteria isolated from the bowels of people living on a high-fat diet and a low-fat diet, as well as differences between those found in people eating a high complex carbohydrate diet compared with a diet low in fibre.

The benefits of a relatively low fat and high complex carbohydrate pattern of eating become clear when the negative health effects of the opposite patterns are looked at, with everything from cancer and diabetes to degenerative cardiovascular and joint diseases being more evident when fat content is high and complex carbohydrates low.

As a third strategy, according to Dr Rašić, in efforts to beneficially influence the microflora [(1) first was use of cultured milk, (2) second was a high complex carbohydrate/low fat diet] we should (3) regularly supplement with bifido and acidophilus bacteria. This can provide a wide range of definite health benefits, as we have seen in earlier chapters. These three elements should be incorporated in any bifidogenic strategy.

If the dietary pattern is difficult to introduce for whatever reason, the addition of cultured milks and supplementation can minimize the dangers of, say, a high meat diet.

COLONIC IRRIGATION

If toxic residues impact in the colon, and if the inhabitants of the region are unbalanced, tending more towards pathogens than friendly flora, it makes sense to wash out (at least partially) the contents and to start repopulation tactics thereafter.

Colonic irrigation does just this, and in the hands of a suitably trained expert, the methods and variations (of solutions contained in the water being used) can make a dramatic impact on restoration of bowel (and therefore of body) heath.

A very large number of disease conditions have been shown to improve after use of colonic irrigation, ranging from arthritis in various forms, to a host of skin conditions, neurological problems and circulatory dysfunction. Many chronic allergy conditions improve or disappear when colonic irrigation is used in conjunction with reformed feeding patterns and restoration of a healthy colonic ecosystem.

More striking still are the reports of the benefits of normalizing the bowel by use of colonic irrigation when treating mental illness. Drew Collins, writing on colon therapy in *The Textbook of Natural Medicine*, describes successful treatment of over 15,000 patients at Gardner State Colony before the Second World War. Dr H. Marshall of that institute stated in *The New England Journal of Medicine*:

'Colon therapy has a real place in the care of mental patients . . . it is rather remarkable that the method has not been used more widely.' Surprisingly, perhaps, even people who report regular daily bowel movements are seen on x-ray examination to retain a large residue of material in the bowel, some of which may have been impacted for years. This may be true even when regular use has been made of laxatives or enemas.

Among the variations of colonic irrigation used, apart from regular tap water, are solutions containing a variety of mineral salts (electrolytes) such as sodium, potassium and chloride. Medications can be, and are, safely administered by this route since absorption is rapid, effectively bypassing the stomach and early digestive processes which would to some extent affect oral intake. Acidophilus and bifidobacteria irrigations are also commonly used to enhance repopulation. This is most useful in cases of candida overgrowth and of general bowel toxicity, and should be accompanied by the oral supplementation which we recommend (see Chapter 17 for dosages.)

Other constituents which may be added to the water used during irrigation include oxygen (an effective means of destroying candida), lemon juice, a variety of herbal substances (fennel, camomile) or Epsom salts. It has been shown in animal studies that repopulation of the bowel after antibiotic treatment, can take up to eight days if left untreated. However, this can be achieved in just two days if oral supplementation of high potency friendly bacteria is introduced.

It has been observed that, during the period of decline of friendly bacteria, after antibiotic use, yeasts rapidly grow onto the lining of the intestines. By using regular colonics (especially if they contain friendly bacteria in solution) during such a period of attempted repopulation, the yeasts can be kept at bay until the friendly bacteria can recolonize the region.

Indications for needing a series of colonic irrigations include chronic constipation, candida overgrowth, bad breath, sick headaches, bloating and belching, arthritis, allergy and food sensitivity, and many of the conditions discussed in previous chapters as being associated with a toxic bowel or disturbed intestinal flora.

Note: All supplementation with friendly bacteria should be with products which carry a potency guarantee of at least one billion viable micro-organisms per gram and such products should carry an expiry date.

A qualified practitioner in the use of colonic irrigation methods can be contacted through either:

Colonics International Association, 26 Sea Road, Boscombe, Bournemouth, Dorset, England.

International Colonic Hydrotherapy Foundation, 62 Alexandra Road, Hemel Hempstead, Herts, England.

Chicago General Health Service, 1614 West Warren Boulevard, Chicago, Illinois 60612, USA.

National College Chiropractic Clinic, 200 East Roosevelt Road, Lombard, Illinois 60146, USA.

Chapter 15

The truth about yogurt and the kefir romance

Cultured milk products are used and enjoyed worldwide. In 1975 the consumption per person per year was an amazing 40 litres. If we consider that some populations ingest hardly any cultured milk at all (the Chinese for example, who prefer fermented soya products) as well as the continued increase in consumption since 1975, the current intake is assumed to exceed a litre per week, for every person on the planet.

In the ten years up to 1979 the consumption of yogurt in America rose by an astonishing 500 per cent, representing a trend which took the volume of yogurt sales in 1986 to in excess of one billion dollars, in the US alone. Unfortunately, with commercial success has come both greed and corner cutting.

Yogurt: the image and reality

For most people yogurt is a convenient (no preparation required) food of bland to sweet taste, with a slight aura of it being 'foreign', and carrying implications that in some way it is 'good' for our health. Research into consumer attitudes shows that yogurt has an image of being 'wholesome, high-protein, convenient, healthy and low in fat'.

To those who may have read something of this 'wonder food' its consumption represents an easy way of contributing towards better health, longevity, and protection against many health problems.

All this is true of real, 'live' yogurt (and kefir, of which more later) 90 per cent of which is digested within the first hour of consumption, as compared with milk of which an average of only 30 per cent is digested within an hour.

These foods, when genuinely cultured and still live, have enormous benefits to offer the digestive processes, including a speeding of the emptying of the stomach and regular healthy elimination. However, the truth is that the vast majority of yogurt sold in the industrialized countries of Europe, and certainly in North America, has little to do with real live yogurt, is often not very good for us, and provides little benefit in terms of health enhancement. It has become in the main a degraded, and in many instances a frankly disgusting, substance, as we will discover later.

Commercial interests have ensured that much so-called yogurt never sees the ideal yogurt-making friendly bacteria, and even if it does do so in the early stages of manufacture (the fermentation phase), these bacteria are usually well and truly dead by the time the product reaches the shelves of most shops.

Most commercial yogurt manufacturers pasteurize their product *after* the culturing process, killing all friendly bacteria (if any have been used) so that the product will have a longer shelf-life (not in New York state though, see below).

If yogurt is indeed cultured with *L. bulgaricus* or *S. thermophilus* (the true yogurt cultures) before these are 'killed' for purely commercial reasons, then the nutritional benefits of virtually predigested protein and milk sugar (lactose) do of course remain. However, no viable health-enhancing friendly bacteria will accompany the food.

Commercial interests have also dictated the corruption of a great deal of this once fine food by the addition of fruit, which if present together with live cultures would result in an explosive mixture, unless of course chemical barriers are provided to come between the fruit (and its inevitable accompaniment of sugar) and such live cultures.

The changes in the way most modern yogurt is produced have been the result of the need to reduce costs, as well as to have a final product which has a shelf-life of some weeks rather than days, and which meets the tastes of an age which has become conditioned to sweetness.

The real yogurt flavour

For the most part then the mildly acid, naturally sweet, taste of real yogurt is no longer found, with sugars, jams, processed fruit and other flavours masking this. Also largely gone in the final product are the live cultures which characterized real yogurt, and with these has gone many of its health-giving potentials, for only if 'live' yogurt is obtained is there a chance of any of the health-giving characteristics of the friendly bacteria being found.

The flavour of live yogurt as created by the activity of selected strains of *L. bulgaricus* and *S. thermophilus*, is usually an indication that there is more L+ lactic acid in the product. This L+ form of lactic acid is more beneficial than D– lactic acid which is produced by less desirable strains and which results in a more tart, unpleasantly acidic taste. D– lactic acid is also produced by strains of *L. bulgaricus* which are used in a number of culturing processes in which there is inadequate attention to such matters as purification and regular checks on bacterial mutation.

The story of real yogurt

It was nearly 100 years ago that Dr E. Metchnikoff first expounded on the link he assumed between the longevity of Bulgarian peasants and their enormous consumption of 'yohourth' (yogurt). The folklore tradition of its usefulness in health maintenance was eventually substantiated by scientific research which showed that it was the lactobacilli involved in the fermentation of milk which gave it these values.

The nutritional value of the milk-base was shown to be enhanced by the fermentation process, making more readily available (and in some instances actually increasing) the proteins, milk sugars and other nutrients (such as calcium).

In addition we became aware of the kaleidoscope of benefits which the live bacteria are able to provide (vitamin synthesis, anticarcinogenic, antitumour, antiviral, antibacterial, antifungal, anticholesterolaemic and detoxification qualities).

These benefits are available when the live cultures are ingested along with the nutritious fermented milk in which they are living. The bacteria which are traditionally used to manufacture yogurt are *Lactobacillus bulgaricus* and *Streptococcus thermophilus*, although the Japanese have produced a fermented milk product called 'yakult' which is cultured with *Lactobacillus casei*. Dr Gilliland of Oklahoma State University reports that this produces similar benefits to those derived from yogurt.

Production: then and now

Historically the fermentation processes which are involved in turning milk into yogurt were unpredictable and slow, depending as they did on micro-organisms inherent in the milk. Modern technology uses either special lactic acid producting organisms (or other acidifying techniques) which are 'inoculated' into the milk under precise pH conditions, in which temperature, water content and other factors are carefully monitored and controlled.

Without doubt this generally does take place in conditions of greater sanitary excellence than in days gone by, and if suitable organisms are used the final product is more physically and chemically stable as a result of modern technological expertise.

We have seen in earlier chapters that particular micro-organisms have specific qualities and abilities which show clearly in their relation to other micro-organisms and their influence on our health status. *L. bulgaricus* and *S. thermophilus* are *the* cultures capable of producing real yogurt. These have been researched to show that they are capable of providing us with a wide range of health benefits, most particularly to do with their bifidogenic qualities. *L. bulgaricus* has of course also the ability to produce its unique antibiotic substances.

Anyone for slime?

However, in manufacturing yogurt, there is no legislation as to which cultures should be used as starter organisms. Often official guidelines and standards will simply list a potpourri of micro-organisms which are acceptable, taking no account of differences in their health potentials.

The main concern of such agencies seems to be the prevention of the presence of patently undesirable organisms: moulds, coliforms and contaminants. This important aspect of cultured milk production is the only area regularly checked. There are no resources for, and little interest in, checking for which starter organisms are used in processing cultured milk into yogurt or other foods.

This abdication of responsibility by many government agencies in giving no direct and clear instructions as to what is and what is not acceptable in cultured dairy products has led to some very strange methods being employed.

Thus one of the major international yogurt producers (nameless, alas, for legal reasons) is not using a lactobacillus starter at all in its fruit flavoured 'yogurt', but a culture called pima. Pima cultures were originally from Finland, and use of these in manufacturing 'yogurt' provides very real advantages to the manufacturer, but certainly not to the consumer.

Pima is a voluminous producer of slime with no known health producing qualities at all. Why, we may justifiably ask, should anyone wish to produce food which is slimy? In normal commercial yogurt production a separation process called 'watering off' takes place, making it necessary for a quantity of milk solids (such as evaporated milk, or powdered or concentrated skimmed milk) to be added, restoring the homogenous constituency of the product. Other substances which some producers add to counteract this normal process are gums (such as carrageen from sea-weed, or guar-gum from beans). Another, and far cheaper, way of avoiding this process is to use a 'slime' producing organism such as pima.

The result is a slimy mixture which is sold as yogurt, but which has almost none of its virtues. In order to mask further the slimy texture of the final product, resulting from the interaction between pima culture and milk, there is usually an addition of processed fruit to the concoction (plus sugar of course) rendering the ingestion of this 'food' a minor health hazard rather than a benefit.

What about fruit flavoured yogurt?

To incorporate fruit into yogurt there is a special procedure that manufacturers follow. The highly processed fruit (or more accurately a jam or jelly) is placed in the yogurt tubs first, followed by a measured amount of cultured milk. The container is then put (along with thousands of others) into an incubation room where it remains for up to eight hours. During this time the fruit has to be kept physically separate from the bacteria in the milk, which is incubating and fermenting, and so a barrier of chemicals is placed above the 'fruit' layer to buffer it. Were this not done the bacteria in the culture would happily start to digest the fruit and its sugar, and would thus neglect the job at hand, of fermenting the milk.

So, people continue to enjoy their tubs of slime and chemicals; although, perhaps with an increasing awareness of what is being so widely marketed as yogurt, many are now trying to find live yogurt. Thankfully, this is still available, more usually in specialized food shops or health stores. If you want to have fruit with your yogurt why not add fresh real fruit at the time you are going to eat it?

Do we get what we are paying for?

In an exercise to discover the accuracy of product description or labelling in America, Dr Gilliland, and his associate Dr Speck, purchased and analysed 12 products from pharmacies, health stores and dairy counters in supermarkets. *All* were labelled as containing *L. acidophilus*.

The first stage of the test was to see whether whatever organisms (if any) were in the products could survive the digestive acids they would meet after being eaten. Out of 12 products only five contained bile-resistant organisms, and only three of these actually contained any *L. acidophilus*. So, less than half of the products assessed in this, admittedly small, survey had any live cultures which might survive the stomach and small intestine, and a mere quarter contained what the label stated was the active culture in the product. Regrettably this state of affairs, of what amounts to dishonesty, abounds in much of the food manufacturing industry, and we are obliged to urge that only the top brand names be trusted, unless a product comes recommended by a health professional or a retailer whom you trust.

Should we eat yogurt and is it safe?

Yogurt is an excellent, virtually predigested food. If it contains live friendly bacteria it is potentially enormously beneficial.

But how do you know if it is alive?

You take a few tablespoonsful of a yogurt which has been marketed as 'live' and mix it with a cupful of regular milk which has been heated (not boiled). Leave this overnight in a warm (not hot) place and in the morning see whether the milk has thickened. If it has then you have indeed bought a real cultured yogurt which has not been sterilized by pasteurization. You may still not be sure which live cultures (bulgaricus etc.) are present and this would take actual biochemical investigation to establish. You may have to rely on the statement of the manufacturer who, if the product does not list this information, can be contacted and asked.

Is yogurt which is not live useless?

Even if bought yogurt is no longer 'live' it has enhanced food value over the milk from which it was prepared. It will contain more readily available protein, calcium and milk sugars than most other daily sources.

However, these advantages are counterbalanced by the fact that almost all commercial yogurts contain stabilizers and a variety of additives. Fruit flavoured ones contain even more chemicals (colouring, flavouring and other substances such as those designed to separate bacteria from fruit during culturing) as well as sweeteners. We urge you to read all food labels carefully and to avoid products which have such chemicals in them, especially if you are sensitive or allergic to foods with chemical additives.

For reasons which will become clear below, we also suggest that extreme scepticism be exercised in regard to any yogurt which claims to contain *L. acidophilus*. In America, some states forbid pasteurization of yogurt after fermentation which means that only live versions are available. New York is one such state.

Make your own

The best of all options, if live yogurt is not readily available from stores, is that you obtain a starter culture which contains the correct ratio of *L. bulgaricus* to *S. thermophilus*, and that you thereafter make your own version of this wonder food. If you wish this to be low in fat you should use skimmed-milk.

The perfect blend of the two yogurt cultures should contain seven parts of *S. thermophilus* to one part of *L. bulgaricus* (depending upon the strains used, for some have more of the recognized health potentials than others and some survive digestion better than others).

The Bulgarians arrived at this ideal symbiotic ratio because it allows both cultures to thrive, and at the same time checks *L. bulgaricus's* tendency to mutate and to create excessive acidity, which can occur if different combinations of quantities are used. Yogurt resulting from this ratio of bacteria is mildly tart, smooth and creamy, with a refreshing aftertaste. We suggest that only such a balanced, tried and tested, combination of yogurt starter bacteria can produce the ideal home-made yogurt with the best nutritional and health-related benefits.

It is worth reminding ourselves that whatever strains of *S. thermophilus* and *L. bulgaricus* are used none can do any harm to us. The very worst that might occur, if weak strains are used instead of the better ones, is that less good would be derived than might have been possible.

We strongly urge you not to add *L. acidophilus* to any home-made yogurt culture as this organism will usually not survive for long in the company of *L. bulgaricus*. The reasons for this are explained below.

If you wish to make your own yogurt you should, for good results, only use a culture-mix which has been specifically designed for this purpose. Do not try to make yogurt by using *L. bulgaricus* or other cultures which are designed for supplementation. Results will be poor.

It is, of course, possible to use live yogurt as a starter culture, as in the 'test' described above, but you may then still not be sure of which organisms are present, or of their ratio to each other.

What about frozen yogurt?

The nutritive value of this popular snack is somewhat tarnished by the addition of vast quantities of either sugar or artificial sweeteners which have dubious health potentials, to put it mildly. The amount of colouring, flavouring and stabilizing chemicals in frozen yogurt is also likely to be large and the quality of milk used in such processes is commonly poor. Home-made iced yogurt is again the best bet.

What of the future for yogurt (and what happens when acidophilus and bulgaricus are put together in yogurt)?

Dr Rašić, acknowledged as a world authority on the subject of cultured dairy foods, says: 'Few consumers realize the importance of strain selection, strain compatibility, ingredient quality and manufacturing procedures on the acceptability of the final product. With advances rapidly being made in genetic engineering of lactic starters, simplification of yogurt manufacturing will probably result, especially to eliminate the need for use of both *S. thermophilus* and *L. bulgaricus* in starters. . . . Enhancement of the therapeutic and healthful properties of yogurt may also occur through wider use of bifidobacteria and *L. acidophilus* in yogurt starters. More research is needed, however, before the latter is possible, especially to develop methods to co-cultivate *S. thermophilus* with bifidobacteria or *L. acidophilus*.'

This last point is important because, when different cultures are mixed haphazardly, there is often excessive growth of some at the expense of others. This happens, for example, when *Streptococcus faecalis* (whom we will meet again later when we consider supplementation of friendly bacteria in Chapter 16) is used together with the better-known friendly bacteria, such as acidophilus and strains of the bifidobacteria.

Japanese research has shown that when *S. faecalis* is cultured together with *L. acidophilus*, the acidophilus will be repressed. This, however, does not happen when *Bifidobacterium bifidum* and *acidophilus* are cultured together, for in this instance the bifidum are repressed. In other cultures faecalis grew abundantly when in the presence of bifidum, although this (bifidum) showed only poor development.

Interestingly, and importantly, for those who wish to have acidophilus in their yogurt, American research by Drs Gilliland and Speck has shown that something strange happens when *L. bulgaricus* and *L. acidophilus* are left together in yogurt.

A steady decline is observed in acidophilus when it is present in a yogurt in which bulgaricus has been used as a culture. The reason for this was eventually discovered to be because bulgaricus was producing hydrogen peroxide (H_2O_2, or common bleach) which was rapidly killing the acidophilus. Some of our own immunological defence cells manufacture minute amounts of hydrogen peroxide as one of the ways in which they immobilize invading micro-organisms. When this happens in the body we have present an enzyme called catalase, which deactivates the bleach before it can do any harm to the normal cells of the body.

Indeed, this is the way in which the method of control exerted by bulgaricus was identified, for when catalase was added to yogurt which contained acidophilus, the ability for bulgaricus to control or destroy it ceased. The marked damage to acidophilus when present together with bulgaricus emphasizes the need for care in selection of yogurts which are commercially marketed, for many of these now seem to carry information stating 'contains acidophilus cultures'. This claim is hard to dispute since the cost of analysing a mixture of *L. bulgaricus* and *L. acidophilus* to establish how many of each are present in viable quantities would be very high and probably impossible.

If such yogurt contains active *L. bulgaricus* (and if it doesn't, then it isn't real yogurt anyway) then because of their antagonism when in close proximity it is sure that the acidophilus content will be negligible.

Other cultured dairy products

The extraordinary story of kefir (rhymes with see-her) deserves retelling. This cultured milk product comes from the Caucasus region of the Soviet Union, where its name has the meaning of 'good feeling' or 'pleasure'. It has long been thought of by the mountain people of its origin as a gift from the Gods. Elsewhere in Europe and in those parts of the US where real kefir is obtainable (New York and Pacific Northwest states mainly) its tremendous nutritional value is also beginning to be appreciated.

In appearance and flavour it is similar to a fermented, slightly alcoholic drink of mare's milk called koumiss. Kefir should contain a blend of cultures including *Saccaromyces kefir, Torula kefir, Lactobacillus caucasicus* (also now known as *L. brevis* and now generally and more correctly described as *Lactobacillus kefir*), *Streptococcus lactis* and others.

Approximately 5 to 10 per cent of the grains of the culture, which looks something like a cauliflower, should be yeast (torula) which imparts its slight alcoholic content (around 3 per cent). The milk used is that of either goats, sheep or cows. The unique flavour is a combination of the lactobacillus acidification and the alcohol production of the yeast.

The story of how kefir was brought out of the Caucasus reads like the storyline of a Hollywood B-movie (one that did not get made, alas). However, it happens to be true.

The cauliflower-like grains of kefir culture were thought of as having amazing healing powers as far back as the eighteenth century, and great care was taken by the Moslem tribesmen of the Caucasus not to allow any of it to pass out of their control, for they feared losing its healing powers. It was virtually bequeathed from generation to generation as a source of family and tribal wealth.

Word of this powerful food/medicine spread to areas far from the Caucasus, and at the beginning of this century the All-Russian Physician's Society asked two brothers named Blandov, who owned cheese manufacturing factories in the Northern Caucasus town of Kislovodsk, for help in obtaining the culture grains of kefir.

One of the brothers, Nikolai Blandov, persuaded a lovely young employee, Irina Sakharova, to use her beauty to try to gain access to the much desired grain. She therefore travelled to the Caucasus where she attempted to interest a local prince, Bek-Mirza Barchorov, to assist her in this plot. When he declined to give up any of the precious substance she left to return to Kislovodsk, only to be captured by agents of the prince, who not content with not giving up any kefir did not wish to lose the presence of the lovely Irina either. Finding herself back in his presence and facing a proposal of marriage into the bargain she remained silent until a rescue mission arranged by her employers freed her. She promptly brought the prince before the Tsar's court where she accepted grains of kefir as the settlement of her suit for abduction.

In September 1908 Irina Sakharova brought the first bottles of kefir to Moscow for sale where it was at first used for medicinal purposes. In 1973, Irina, then 85, was sent a letter from the Minister of the Food Industry of the Soviet Union, acknowledging her great part in bringing kefir to the Russian people.

It was one of us (NT) who in the early 1960s obtained from the Soviet Union grains of kefir which were brought to the US. Unfortunately the Californian authorities ruled that cultured dairy products should not contain yeasts, and so the kefir obtainable in that state is a modified version of the traditional drink.

This barrier does not exist in other parts of the country (although for reasons of taste preference as well as practical ones such as shelf-life the New York kefir is lower in yeast content than European brands) and it is to be hoped that a standard can be set based on the New York guidelines which will allow the rest of the country to enjoy the benefits and taste of this marvellous food. A hopeful indication that this may happen is that some torula-based kefir is now available in the Pacific Northwest of the US.

Why does it matter? Because scientific evidence of the effects of fermented milk products, including acidophilus milk, acidophilus yeast milk and kefir in the USSR (by I.V. Kasienenko) shows that kefir far surpasses the others in its benefits, including much enhanced stomach secretions and a higher nutritional content of B vitamins and other essential substances. As with yogurt, so with kefir, and indeed all fermented dairy products, the key to their useful, widespread use lies in the establishment of realistic standards of manufacture. Until then buyers need to become aware of shortcuts taken and of precisely what to look for when they buy what can be a marvellous health food but which is all too often not.

This is true of cottage cheese, acidophilus milk, koumiss, kefir and yogurt and, as we will see in the next chapter, it is equally true of those friendly bacteria we can supplement as preventive and therapeutic agents.

Chapter 16

The supplementation maze: how to get the right products

The good news

The facts about the multiple benefits available from the friendly bacteria have now been laid before you. Their ability to help your body in so many ways is fact, not fiction, and you can further enhance this by means of a dietary approach which was explained in Chapter 14. You can also use a more direct method of enhancement of the friendly bacteria, that of providing them to your body in the form of food supplements.

By supplementing the major role players themselves, *Lactobacillus acidophilus* and/or bifidobacteria, you improve the chances of increasing the numbers of these which will attach to their normal sites in the GI tract and begin to do their amazing work.

Supplementation of resident bacteria (*L. acidophilus*, bifidobacteria) can be critically important after antibiotics have been used or after any bowel dysfunction such as diarrhoea. It is also a useful strategy to replace or enhance them by supplementation in a host of acute and chronic health problems, as has been discussed in previous chapters.

By supplementing associated transient (non-resident) bifidogenic bacteria, such as *Lactobacillus bulgaricus* or *Streptococcus thermophilus*, you can further aid in the health and function of the resident friendly bacteria. The really good news is that despite some of the extremely worrying facts which we will present in this chapter regarding the sometimes atrocious, and frequently useless, acidophilus-type products on the market, there are excellent ones as well.

We will help you to identify both the good and the bad.

The bad news

In order to be of any real value and to be able to take up residence in the GI tract (or to enhance the resident friendly bacteria) any supplemented cultures need to contain very large numbers indeed of suitably viable organisms, and they need to be made up of strains which will happily survive the digestive process.

Over the past few years, as more and more people have become aware of the usefulness of supplementation with friendly bacteria, especially of certain strains of acidophilus, a marketing phenomenon has become evident. This can be termed the 'somethingdophilus' phenomenon. There are now so many 'dophiluses' in health stores and pharmacies, not to mention health food racks in supermarkets, that the public may justifiably become confused as to which are sound and which are not. The price variations between these various 'somethingdophiluses' is enough to confuse and depress the potential purchaser, ranging as they do from fairly cheap to what appears to be exorbitant (see the end of this chapter for comments on price).

This confusion is compounded by the label notes which often tell the buyer that the content possess tens, or hundreds, of millions or even billions of active organisms per gram. How many do we need to do any good anyway? And more recently, to add yet further to the confusion, there have arrived a range of bacterial cocktails, usually containing a certain 'dophilus' content as well as a variety of other organisms (such as *Streptococcus faecium* and *Lactobacillus casei*) almost always in unspecified quantities, except that the word 'billion' usually appears somewhere on the label.

Questions which need to be asked

There is a maze of confusion and, unfortunately, in some instances a good deal of deception in the commercial world associated with the friendly (and sometimes not so friendly) bacteria. Vested interests have come into play and this has not always been to the advantage of the consumer who should be formulating and asking some pertinent questions of the supplier or manufacturer, as to the quality, reliability and effectiveness of what is being sold.

What is that we need to ask?

In the following pages we have framed what we consider to

be important questions and have attempted to answer these objectively. In doing so we have relied to a large extent on the enormous body of published research on this subject, references for which are given at the end of the book.

We also rely on the extensive and intimate knowledge of this subject which each of us possesses, with NT, having spent more than 25 years in research into this subject alone, being generally acknowledged as a major authority in the field of cultured milk micro-organisms, and L.C. having spent an equal amount of time involved in the treatment of human illness, as well as in health education and health enhancement, using nutritional (and other non-invasive) means, including the supplementation of bacterial cultures.

We suggest that you read these questions and study the answers, referring to the earlier part of the book for assistance in clarifying aspects of the answers.

This is an important exercise and one which could make the difference between your gaining health benefits or not, as well as possibly helping to save you from wasting hard-earned money on undesirable imitations of some of the truly excellent supplements which are available. In truth the best friendly bacteria supplements are so good that it does pay to seek them out. The worst are, we believe, useless and sometimes, frankly, dangerous.

We urge you to ask the following questions of those who sell 'somethingdophilus' products, and to compare the answers you get with the information and explanations we have provided.

Questions and answers

1. How many viable organisms, of any of the friendly bacteria, should there be per gram of such supplements, if there is to be any health benefit?

In order for a therapeutic effect to be possible the number of viable colony-forming organisms needs to be in the billions per gram. This allows for some loss in their passage to the small or large intestine, whichever is their final destination (or in the case of the transient bacteria, of their passage through the GI tract).

The most important fact which you need to be aware of here is that the numbers should be of *viable* organisms, not just fragments of these. This will to a large extent depend upon the means used to separate the bacteria from their growing medium, i.e. ultrafiltration or centrifuging. This most important subject is discussed in detail in question 3.

Supplements which state their content in millions of organisms per gram are probably not adequate for therapeutic use. They may, however, help somewhat in maintenance of an already healthy bowel ecology, provided that the organisms present in the supplement are of suitable strains and are viable and capable of colonization at the time of purchase and right up to expiry date.

We suggest that, if all other elements (see questions below relating to centrifuging, different strains and mixtures of organisms etc.) are satisfactory, you should always aim to take bacterial supplements which contain *several billion* live, viable, organisms per gram.

Note: In the first draft of a proposed National Nutritional Foods Association (USA) labelling standard for probiotic products it states: 'In any product which uses the expression "dophilus" in its name, at least 50% of the viable organisms present at the expiration date shall be *Lactobacillus acidophilus*.' This standard is not always being met.

2. Are liquid acidophilus products a waste of money?

Most health food consumers who purchased liquid acidophilus products some years ago continue to buy the liquids today, because they may be unaware of its instability and rapid potency deterioration. Many consumers are buying these liquids because of price considerations. However, unless the manufacturer guarantees the potency at the time of purchase, through to the expiry date, the consumer is getting a nutritional residue of *L. acidophilus* and lactic acid, but virtually no living active acidophilus cells.

Possibly a better name for liquid acidophilus would be 'nutritional by-products of *L. acidophilus*'. There is some nutritional value, but not in comparison with the billions of active organisms available from powdered products.

3. Can you trust the potency numbers given on the label; how was the product cultured, and was it centrifuged before this count?

Both the public and retailers have been inundated with information regarding the potency of different friendly bacteria products. Millions and billions of micro-organisms per gram are quoted, with numbers rising as each new product reaches the market.

The truth is that not all producers have counted the same things. It is as though, in counting bricks for us in building a wall, a builder were to count all the fragments of broken bricks as thought they were whole ones. In order to understand how this analogy could possibly be applied to friendly bacteria we need to look briefly at the process of their manufacture.

Bulk manufacture of the various cultures starts (usually — see question 6) with a milk medium since the bacteria readily thrive in such an environment.

Unfortunately, some manufacturers then add booster chemicals, and sometimes even hormones, to stimulate growth of the bacteria. These do not have to be listed on the product label since they are regarded as 'processing agents' and not ingredients. Once a suitable degree of growth has occurred and in order to rapidly separate out the bacteria from the medium in which they are growing some manufacturers then add more chemicals, and introduce a rapid spinning procedure which can have disastrous effects on the fragile bacterial chains.

Acidophilus, for example, normally grows in long chains which are called 'colony forming units'. If these cultures are spun at very high speed during the centrifuging process, which separates them from their growing medium, the delicate chains are likely to fracture, reducing the final result to a collection of links from the original chain (or broken bricks). Not only are the chains fragmented, but individual cells become damaged due to the violent G-force (and the chemicals) to which they have been exposed.

If the intact bacterial chains (un-centrifuged, that is) are examined, each chain represents one potential colonizing unit, which could be expected to be able to provide therapeutic benefits, since each viable chain of acidophilus is capable of producing up to 20 new colony 'families' once suitably implanted in the GI tract. Each complete chain is counted as one organism.

If, however, after centrifuging, the individual links of fragmented chains, and naked cells, are counted, we would certainly arrive at a higher number, but in this case the organisms counted would not represent viable colonizing units, and they would have little chance of delivering major benefits, since such individual particles of the splintered chains (each counted as a viable unit, remember) usually result in no more than three new colony families each.

Thus, in un-centrifuged bacterial counts you may see that there are several billion viable organisms, each capable of creating up to 20 families. On the other hand, if the product is centrifuged there may also be several billion organisms present, but only capable (at best) of producing three families each.

For this reason alone, it is quite obvious which product we should choose when faced with either centrifuged or un-centrifuged supplements. Unfortunately that is not all, for over and above the fragmentation of viable chains, something equally serious occurs when premature centrifuging of this sort is performed (and this is all too common in the manufacture of 'somethingdophilus'). After centrifugation the acidophilus organisms will have lost their natural original milk medium containing acidophilin, the natural antibiotic which specific strains of this organism produce. Remember that it is largely through its ability to create acidophilin that ascidophilus is so effectively able to control pathogenic (disease-causing) micro-organisms.

Note: In the process of ultra filtration, the cell walls are not damaged as they are in centrifugation. Both methods, however, remove the original growth medium which contains very important and beneficial by-products of the bacteria's growth. In Chapter 5 we talked about Dr Ninkaya's research which clearly stresses the importance of the supernatent (original culture medium).

UK Comparative study

In 1985 one of us (LC), in his capacity as Consultant Editor of *The Journal of Alternative and Complementary Medicine*, initiated an analysis of three of the most popular branded acidophilus products then on sale in the UK (all were of American origin) in order to assess the accuracy of their claimed potencies.

The results were published in the March 1985 issue of that journal (pages 20–21) and were of considerable interest. The products assayed (independently, by acceptable scientific methods) were Superdophilus, a powder manufactured by Natren Inc. of California and marketed in the US as Megadophilus; Vitaldophilus, also a powder, manufactured by Klaire Laboratories; and Natural Flow, encapsulated powder.

Superdophilus (which requires refrigeration before and after opening of the container) claimed to contain not less than a billion viable organisms per gram. The assay found 8 billion viable organisms per gram.

Vitaldophilus (which requires refrigeration only after opening of the container) claimed 10 billion organisms per gram. The assay showed 2 billion viable organisms per gram.

Natural Flow made no claim, but assayed as having at least 1 billion organisms per gram.

A popular brand of acidophilus tablet was also assayed and was shown to contain less than ten million viable organisms per gram.

These results highlight the need for caution over claims. In fact, all of the products assayed (apart from the tablets) would appear to be able to provide benefits, having potencies of at least a billion organisms per gram. However, only the manufacturers of Superdophilus stated clearly that their product was un-centrifuged.

It is, we contend, particularly important to look for the information 'not centrifuged' on labels of acidophilus products. If no such assurance is forthcoming, ask the manufacturer directly whether this process is used in their production methodology. Ask also whether the count ('so many million or billion organisms per gram') given on the container, or in the advertising of a product, represents intact viable chains of bacteria or fragments of spun cultures, and whether this count represents the numbers present at the time of manufacture of that anticipated at the time of purchase, and through to the expiry date on the label.

If you fail to take these precautions you may end up with a product with seemingly large numbers of organisms which are actually only fragments of useful material, not possessing the major ability to actively re-colonize the GI tract or to produce natural antibiotic substances for your (and its own) protection.

Unfortunately, some manufacturers provide a count of micro-organisms based on the number assayed at the time of production and unless there is extreme caution taken by them over factors such as temperature control during handling, delivery and storage, most potency claims probably represent no more than wishful thinking as to what will be present at the time of purchase and through to the expiry date.

It is also important that you discover whether the claimed numbers of micro-organisms per gram are verified by an independent laboratory assay, each time a new batch is produced. Only if you are sure that each batch is accurately and independently assayed (that is not by the manufacturers themselves) and that the product is not centrifuged, can you be certain of a reasonable chance of therapeutic or preventive activity from the friendly bacteria.

We wish to make it clear that we know that many suppliers of cultured supplements are quite unaware of many of the facts we have explained in this section. This is because very few suppliers of branded supplements actually manufacture their own products. In the main they buy these in from major manufacturers, and trust that what they are getting (and selling) bears a close relationship to what they are advertising.

We hope that the information we have given regarding fragmented chains of chemically denatured organisms with minimal health potentials will give some suppliers and retailers reason to pause and reconsider their marketing strategies and claims.

If major suppliers, with good brand names, can pressure the manufacturers to alter their methods we should see a revolution in the quality of 'dophiluses' reaching the public.

Not all supplements of this sort are denatured, of course. Some manufacturers specifically state that their products are not centrifuged, and that they obtain independent assays of each new batch, proving their claimed potency figures. We urge you to stick to such products and to let the retailer from whom you purchase your supplements know the reasons for this choice.

4. Should the individual strains of the various friendly bacteria be named on the container, and does this really matter?

We have seen in earlier chapters that some strains of *L. acidophilus, L. bulgaricus* and the bifidobacteria have unique qualities not shared by the fellow members of their kind.

For example, LB-51 has strong anticarcinogenic and antitumour properties and is the most prolific producer of lactic acid. It hardly pays to buy a cart-horse if a race-horse is required. Unfortunately we cannot tell simply by looking which bacterial 'horse' is in any given product.

If information on labels, or in advertising, of particular supplements of the friendly bacteria provides the name of the strain, this gives you the chance to establish whether this strain is desirable or not. It also indicates that the manufacturer has a degree of confidence in their product. You should certainly try to ascertain which strains are being used if this information is not provided.

Proven super-strains

Strain DDS-1 of the acidophilus group is truly a super-strain which has been widely researched and its abilities documented by scientists such as Dr Khem Shahani at the University of Nebraska. It has powerful anticarcinogenic properties and (among many other capabilities) produces effective natural antibiotic substances which can deactivate eleven known disease causing bacteria (and 16 harmless ones).

There are 200 different strains of *L. acidophilus*, some 13 of which have quite strong antibiotic qualities but DDS-1 was shown by Dr Shahani and his colleagues to be the most effective against pathogens. Other acidophilus strains available commercially are not always designated by name or number. The only other identified strain available in the US is the NCFM strain which is widely used in commercially available 'sweet acidophilus milk' which does not contain large numbers of this strain (usually no more than one or two million organisms per gram). The NCFM strain does not inhibit as many disease causing organisms as does DDS-1

Bulgarian research over the past 20 years, mainly by Dr Ivan Bogdanov, has shown that the most effective strain of *L. bulgaricus* is provided by LB-51. This transient organism makes a positive contribution to the viability of the resident friendly bacteria, assisting in their growth and activity.

Another strain of *L. bulgaricus* which has strong antibiotic and anticarcinogenic properties is DDS-14, but this is seldom used commercially. Supplementation of bifidobacteria, especially to babies, should be either *Bifidobacterium infantis* (for the very young) or *Bifidobacterium bifidum* (strain Malyoth) for the older child and adult. Yes, it does matter which cultures you supplement, and it also matters which strains of those cultures you use, if the best results are desired.

5. Does it matter what sort of container the culture is sold in, glass or plastic?
We believe, for a variety of reasons, that the ideal packaging material for cultures is amber-coloured glass which has been sterilized just prior to filling.

Such a coloured glass container prevents access to the cultures of light and moisture, which could diminish the potency of the product. The porosity of glass is negligible, whereas the least porous plastic is too porous for *L. acidophilus* or other bacterial products. The worst possible scenario is the manufacturer who packs his *L. acidophilus* in plastic and then asks the retailer to refrigerate it. Since most retailers do not maintain their refrigeration units properly, humidity inside is very high. The porosity of plastic allows moisture to enter the product causing great damage. In this way an attempt at protection ends as an assult on the integrity of the product. Use glass.

6. Does it matter what medium the cultures are grown in, for example, if a person is milk sensitive will acidophilus grown in a milk base upset them?
Most people on this planet are sensitive, or are actually allergic to, one or other of the contents of milk which has been derived from animals, other than that of their own species (i.e. human).

It has been estimated that 80 per cent of people world-wide are sensitive to either lactose (milk sugar), casein (milk protein) or milk fat (cow's milk). Those people sensitive to lactose can often overcome this by supplementing with *L. acidophilus* DDS-1 which produces copious amounts of the enzyme lactase in which these people are usually deficient. Dosage suggestion for lactose intolerant people are given in the next chapter.

Anyone who is extremely sensitive to milk, however, might find it impossible to take a dairy-based supplement of this kind, even with DDS-1 supplementation. For this reason some manufacturers have created milk-free acidophilus products.

In such cases the growing medium for *L. acidophilus* may be carrots, rice, potatoes or any other suitable foods for the bacteria which would not trigger an allergic response in a milk-sensitive person. Usually a combination of ingredients is necessary to provide a complete diet for the acidophilus, because no one vegetable can provide a complete and balanced diet, such as milk does. Therefore, if a dairy-free acidophilus uses only one vegetable, such as carrots, as a growth medium it would be an incomplete diet and unlikely to maintain healthy viable *L. acidophilus*.

A milk-free acidophilus product would state this fact clearly on the label of the container. All other cautions apply to milk-free cultures, relating to centrifuging etc.

7. What mixtures of cultures are desirable and which not (e.g. S. faecium *and others) in the same container, and does it matter if you take a variety of friendly bacteria from separate containers at the same time (acidophilus, bulgaricus, bifidobacteria etc.)?* We have shown in the previous chapter that when combined together in yogurt, for example, *L. acidophilus* and *L. bulgaricus* will not happily cohabit. In fact, in such a relationship, the acidophilus organisms would be deactivated, leading ultimately to only viable bulgaricus organisms being present.

Although cultures destined for supplementation are usually freeze dried (and may be required to be kept refrigerated) this does not completely stop their activity: they continue to function, albeit in slow motion. There would be competition for food, were different organisms mixed together in a container, and ultimately one would dominate, resulting in an unbalanced product.

We believe that pure cultures are safest and this ensures that there is no such exposure to the danger of combat and rivalry between micro-organisms.

The faecium and casei story: A recent development in marketing of 'somethingdophilus' products has been the emergence of *Streptococcus faecium* in many brands.

Some strains of this bacteria have been widely used in Europe as a means of treating farm animals. They have proved to effectively enhance growth, in that animals supplemented with

S. faecium gain weight faster than unsupplemented animals, which is obviously more profitable for the farmer. *Lactobacillus acidophilus* has similar effects, but is not as hardy a bacteria as *S. faecium* when exposed to the vagaries of heat, moisture and other hostile environmental changes.

S. faecium is one of a group known as enterococci, and these include some potentialy hostile organisms, although as yet there is no specific evidence that the *S. faecium* strains used in human supplementation (usually No. 68 and No. 74) are involved in food-poisoning or other toxic reactions, *S. faecium* has been shown to produce ammonia from the amino acid arginine, and does not have the ability to reduce nitrates as do the major friendly bacteria of the lower bowel, the bifidobacteria.

One of the main reasons for the widespread use nowadays of *S. faecium* is that it is cheap and easy to produce, unlike the more stringent requirements involved in *L. acidophilus* production. This consideration alone, however, does not explain the reason why a bacterial culture usually associated with animal feed has very recently become a major part of an important American health food product (and by extension this applies to the UK and Europe since the major source of brands in the UK and Europe derive from the USA). As mentioned above, most brand names of 'somethingdophilus' and other friendly bacteria culture supplements are not manufactured by the company marketing them. A great many buy in their culture and either pack this themselves or have even that task performed by outside firms, simply confining their role to marketing the product(s).

For some years one of the major bulk dairy-food producers, supplying the majority of 'own-label somethingdophiluses' to dozens of different firms, has been selling them cultures, for use in such brands, of *L. casei*, the main cheese starter culture.

Now, we have seen in earlier chapters that some strains of *L. casei* do have therapeutic potentials, being used in the production of the Japanese yogurt look-alike, yakult. However *L. casei* is not known to produce natural antibiotics and can provide few of the many benefits associated with the bifidobacteria or *L. acidophilus* strain DDS-1.

It may be that *L. casei*, through its ability to acidify the GI tract could be termed a useful friendly bacteria. However, this is not the same as saying that it could seriously be thought of as a suitable replacement for acidophilus, bulgaricus or bifidobacteria.

Nevertheless, for reasons of commercial expediency, for many years bulk dairy suppliers have been providing *L. casei* to firms who have knowingly or unknowingly marketed this as 'somethingdophilus'. Potency (how many million/billion organisms etc. per gram) was seldom specified, and in many instances no instructions were given on labels as to refrigeration of such products.

In 1986 one of us (NT) prevailed upon a major American health food association to assay most of the major acidophilus products on sale in America. The results were so shocking that they were never published. Most were found to contain *L. casei* instead of acidophilus. The shock-waves caused by this potential scandal led to urgent consultation between the major bulk suppliers and the various firms marketing brand named 'dophiluses'. The solution they came up with was the introduction of a cocktail mixture of micro-organisms which included *L. faecium, L. acidophilus*, bifidobacteria, *L. bulgaricus* and *S. casei*.

Suddenly there appeared a rash of new labels informing the public that *S. faecium* was a desirable co-factor organism which should be supplemented in such a cocktail. These are, after all, food supplements, not medicines, and as such their contents should consist of normal bacterial inhabitants of the GI tract, or at least be normal transient visitors such as those organisms found in cultured milk products.

We should also realize that no research has been done to establish that such a cocktail provides a stable environment for any of the bacteria mixed together in it, nor is it clear which bacterial type or strain would predominate in such a situation, for some will always thrive at the expense of others.

The Californian Department of Food and Agriculture, replying to an enquiry from one of us (NT) as to the safety of *S. faecium*, in August 1988, stated:

'[Our] Milk and Dairy Foods Control Branch would not classify these (*S. faecium, S. faecalis*) as pathogenic [disease-causing] or harmful organisms. However, the presence of these organisms [in milk products] would be objectionable as they would contribute to a product with a limited shelf-life and their presence would also indicate possible avenues for product contamination by other organisms.'

S. faecium seems to present little obvious or immediate danger, unless it were to mutate (which is not uncommon in enterobacteria) into a harmful form.

Equally certain, however, is that it possesses few of the qualities of the important and well-researched acidophilus and bulgaricus strains we have discussed above. If manufacturers wish to market *S. faecium* supplements they should label them as such and not trade on the public's awareness of *L. acidophilus* by calling their product 'somethingdophilus'.

Indications are that, because of its hardier characteristics (but not its therapeutic value), *S. faecium* would probably dominate acidophilus and bulgaricus, in any mix of these organisms, leaving a less than desirable end-product for consumption by an ill-informed public.

There is no guarantee as to what percentage of any such mixture of bacteria even starts out as acidophilus, which is surely an undesirable thought if we are trying to gain access to these valuable friendly bacteria.

Throughout this horror story some firms have remained dedicated to providing pure cultures, thus assuring that those food supplements which contain friendly bacteria are of pure content and high potency.

> We once again urge you, if you are contemplating supplementation with friendly bacteria, to ensure that it is just such a brand that you turn to. Always look for statements as to contents being of a single organism, and that this is a strain which you recognize from the information above to be a useful one, that there is no added *S. faecium* or *L. casei*, and that the other elements of importance, as outlined in this chapter (type of container, centrifuged or not, etc.) are met.

There are indeed excellent products on the market, but there are a great many which are at best useless.

When it comes to supplementing with friendly bacteria we should recall that their friendliness extends to us but not necessarily to each other. Therefore, while it is fine to take acidophilus and bifidobacteria at the same time, we should separate the taking of acidophilus and bulgaricus by some hours; that is, if we happen to be supplementing with both. Guidelines on such tactics are given in the next chapter.

8. Does it matter whether the products are refrigerated when you purchase them and do you keep them refrigerated after opening? In the survey made in 1985 (see above) of three popular American acidophilus products we noted that one (Superdophilus) contained some 8 billion organisms per gram, whilst only 'guaranteeing' one billion. Another product (Vitaldophilus) claimed 10 billion and showed 2 billion (still a useful number) per gram.

When we consider this study we should note that Superdophilus is kept refrigerated from the time of manufacture until it reaches the customer (who is advised to keep it refrigerated as well). Vitaldophilus, on the other hand, is not refrigerated until after it has been opened. Is this the reason for the difference between claimed numbers and actual numbers?

Were there in fact 10 billion viable organisms present at the time of manufacture of Vitaldophilus, reducing to 2 billion at the time of its opening?

Natren the manufacturers of Superdophilus have this to say on the question of refrigeration:

'It is common in the industry to claim that refrigeration of a particular *L. acidophilus* product is unnecessary. This practice may be intended to create a false illusion of product stability. In fact, Superdophilus is the most stable product available, yet refrigeration is required on site, in transit, and at the vendor's premises at all times. Indeed, while refrigerated, Superdophilus is potency-guaranteed to have billions of viable *L. acidophilus* up to a year from the date on the bottle.'

We believe that the best products should be refrigerated from site of manufacture to the hands of the purchaser, and this is assured by some companies, even to the extent of having any of the product which is mailed placed in dry ice containers.

If potency is important, and it is one of the major elements in assuring the effectiveness of friendly bacteria as a food supplement, then refrigeration is essential.

9. Is there any difference between bacterial cultures which are marketed in powder form and those sold in capsules (or as tablets)? Encapsulation of acidophilus powders has often been promoted as a means of effective delivery, since capsules can assure greater degree of passage through the early digestive hazards. This may well be true. The problem seems to be that spoilage of the culture seems to be more rapid when encapsulation is used. Independent testing laboratories have almost always found that the potency (number of viable organisms) of encapsulated acidophilus products is well below the billions claimed on the labels after only a short time on the retailer's shelf.

The manufacturing of tableted friendly bacteria exposes these to stresses, damage and contamination from a variety of chemicals and excipients used in such processes. Their viability is usually seriously compromised and we do not recommend such forms of supplementation of these organisms. The burden of proof regarding the numbers of viable organisms which can be assumed to reach the intestines lies with the individual manufacturers of tableted or encapsulated acidophilus supplements.

Some enteric coated capsules presumably can assure a greater degree of passage through early digestive hazards. Enteric coated capsules may allow *L. acidophilus* to continue their life-cycle consuming any available food in the capsule, possibly dying of starvation before they reach you, the consumer.

The only way you can be certain that you are getting value from acidophilus capsules is if the manufacturer guarantees potency right up to expiry date of the product, and that this potency is verified by periodic testing by an independent laboratory.

We, as of this writing, recommend powdered forms as being the most stable way in which you can obtain these important food supplements.

10. Does it matter when you take these cultures, with meals, away from meals, with hot or cold drinks, etc.?
One of the greatest dangers facing most friendly bacteria before they reach their safe haven in the GI tract is passage through the highly acid (less than pH 3.0, usually) medium of gastric juices.

Much depends upon how acidic the stomach is; how long the bacteria remain in the stomach (this depends upon stomach emptying time) and any physical protection the organisms may derive from other contents in the stomach. Much also depends upon the vulnerability of the organisms, a factor which is increased if they have been centrifuged, damaging their protective coating.

If relatively small numbers of damaged micro-organisms reach a very acid stomach (especially at mealtimes) and stay a while they will effectively be destroyed.

If a large number of relatively well-protected (undamaged) organisms reach the stomach when it is empty they will pass through swiftly and many will emerge intact in the small and then the large intestine. This is more likely if the organisms have been cultured on milk which provides them with a natural buffer from the acid.

Those companies which suggest taking friendly bacteria with food seem to believe that the food ingested at the same time serves as some sort of food source for the bacteria, which will protect them from stomach acids. This reasoning is unsupported by any published scientific data.

All the evidence points to the best time for taking such supplements as being an hour before mealtimes. We suggest that they be taken with a tumbler of unchilled filtered or spring water, not with juices, milk or any other beverage which would stimulate production of digestive acids. Instructions for taking supplements will be given in the next chapter. *L. bulgaricus* is best taken with meals (see next chapter).

11. Why are some acidophilus-type supplements apparently expensive and others not?

Most of the better cultures available in supplement form are relatively expensive. This, as we have explained in answer to earlier questions, is largely because of the stringent controls required for the production of high-quality, potent cultures, which require very careful handling at all stages of their production. Quality control does not come cheap. In the case of the better products, as explained above, refrigeration at all stages, during and after production, up to and beyond the moment of purchase, is the ideal. Ensuring this is costly. It is also expensive to provide the detailed care regarding materials used in packaging, a factor which, if not considered, could result in the product losing its viability and potency.

There are certainly many less expensive products on the market, but if their content is often valueless, or close to that, we would ask which is the really expensive item — the cheap relatively useless one or the expensive but useful one?

Chapter 17

Guidelines to supplementing friendly bacteria (how, when and how much)

General purpose supplementation

Lactobacillus acidophilus, in high potency (in excess of one billion micro-organisms per gram) should be taken, in doses of 1 gram daily, indefinitely. This represents approximately half a level teaspoonful of powder, which should be stirred into unchilled spring or filtered water and drunk immediately, three quarters of an hour before a meal.

Ideally this should, for maintenance purposes, be accompanied by bifidobacteria such as the strain *B. bifidum* 'Malyoth' strain, in a ratio of 4 to 1. This means that if half a teaspoonful of acidophilus is being taken, an eighth of a teaspoonful of bifidobacteria powder should be taken at the same time. The 4 to 1 ratio is applicable to people on a 'Western' type diet including red meat. Vegetarians and lacto-vegetarians and athletes should take bifidobacteria supplements exclusively or in a ratio of 4:1 (bifidobacteria to *L. acidophilus*). These bacteria are compatible with each other. *Bifidobacterium bifidum* 'Malyoth' strain is for infants, older children and adults, whereas *B. infantis* is for children under 7 years of age only. A higher ratio (i.e., equal parts bifidobacteria to acidophilus) may be called for if these are being used in therapeutic settings.

In order to enhance the resident friendly bacteria for general health purposes (non-specific), supplement daily with a half teaspoonful (one gram) of *Lactobacillus bulgaricus*. This should be taken *with or immediately after meals* in a juice, milk or mixed with soft cool foods. Specific strains such as LB-51 are triggered into releasing specific health enhancing substances if taken with food.

Therapeutic dosage

If a particular health problem (such as *Candida albicans* or high cholesterol levels) requires supplementation of *Lactobacillus acidophilus* then, along with all other useful strategies (see Chapter 8), take between 5 and 10 grams daily of high potency acidophilus for a week or longer, followed by a reduced maintenance dosage (half the therapeutic intake, perhaps). This is equivalent to 2½ to 5 level teaspoonsful of the powder per day.

The doses should be divided evenly between each meal of the day when high levels are being taken. Ideally, we suggest that this degree of supplementation should be directed or guided by a health care professional. In addition, an equal dosage of bifidobacteria ('Malyoth') will complement the therapeutic use of acidophilus.

Supplementation of *Lactobacillus bulgaricus* to enhance the resident friendly bacteria, thereby controlling pathogenic bacteria and assisting immune function, in a therapeutic setting, requires the taking of between 1½ and 3 teaspoonsful (3 to 6g) with each meal (3 times daily).

If there is a need for liver detoxification this indicates a need to supplement with bifidobacteria strains such as 'Malyoth'. Take these up to 3 teaspoonsful (6g) daily in water 45 minutes before meals.

Children

Children should take reduced doses which takes account of smaller bodysize. Thus, a 35 kilogram (5½ stone) child should take half the quantity recommended for an adult (as above), i.e. for maintenance a quarter of a teaspoonful of high potency acidophilus powder (½ g) daily in water before meals.

In infants, supplement only with *B. infantis* unless there has been a recent infection (diarrhoea etc.) in which case also use acidophilus (50/50 ratio). This is also the ideal supplement for non-breast-fed babies who seldom have high levels of this important organism. (Please see caution about use with infants on page 152.)

Dosage should start at about a third of a teaspoonful of *B. infantis* daily, rising gradually to a teaspoonful in water in divided doses. Some children benefit from the use of both *B. infantis* and the 'Malyoth' (which is resident in adults and children) strain of bifidobacteria, as well as acidophilus (a third of a teaspoonful of each, mixed in water) daily.

Lactobacillus bulgaricus should be supplemented to infants and very young children only under guidance from a health care professional.

Pregnant and nursing women

Enhancement of the presence of *B. infantis* is very important, as explained in Chapter 13. Women expecting or nursing a baby should take *Bifidobacterium infantis* in doses of 1 to 2g (½ to 1 teaspoonful) daily in water before meals.

Milk intolerance

If you are intolerant to milk or milk products, either by manifesting clear allergic symptoms or mild digestive discomfort after consuming such foods you should be able to overcome this, especially if the sensitivity relates to the lactose (milk sugar) component of the milk, by using small quantities of *very high potency acidophilus powder (strain DDS-1)*.

Begin with an eighth of a teaspoonful of an acidophilus which has been cultured on milk and which has not been centrifuged, and which contains not less than 1 billion organisms per gram. Take this quantity in spring or filtered water (tepid, not cold) once a day about three quarters of an hour before a meal, for three days. If there is no adverse reaction (milk allergy symptoms) increase this to a quarter teaspoonful of powder in water, three quarters of an hour before a meal, for another three days. If no adverse reaction occurs to this, increase the dosage, then start to take a quarter teaspoonful of powder in water twice daily for another three days.

Continue to increase the quantities thereafter so that you are taking a half teaspoonful, either one or three times daily. The variations will depend upon the quantity which you require to prevent allergic symptoms from manifesting.

If there is poor tolerance of milk-based acidophilus, when small amounts are attempted as described above, take instead a *dairy-free version of acidophilus*.

How to avoid travellers 'tummy' (and what to do if it happens anyway)

Take a *high potency acidophilus* powder with you on your travels, keeping the container refrigerated as far as is possible. It can be kept in a cool-bag or well wrapped in aluminium foil for those parts of your trip when you are away from refrigeration facilities. On board planes and boats, and in many hotels, refrigeration is usually available for such needs. If it is not, keep the container in the coolest place you can.

Two weeks prior to departure start to take half to one teaspoonful (level, not heaped) of high potency acidophilus (strain DDS-1 which produces acidophilin) in 8oz of filtered or spring water (large tumbler) three quarters of an hour before mealtimes, two or three times daily.

While on holiday, continue to take between half and one teaspoonful of powder 45 minutes before a meal, at least twice daily, in water. If there is any sign of food poisoning (vomiting, diarrhoea etc.) increase dosage to a heaped teaspoonful immediately and continue to take this amount frequently (every few hours) until the problem abates. Take plenty of liquids and call a doctor (use of such supplements is not meant to replace expert medical attention).

If the product you purchase is uncentrifuged it will not become unduly damaged if it is exposed to brief periods at high temperatures. However, try to ensure it is cool, or better still refrigerated, as far as is possible.

A chart of micro-organisms used as probiotics (compiled by J. Lj. Rašić Ph.D.)

1. *Lactobacillus kefir*	A transient bacteria of the human intestine. Found in kefir grains and kefir drink. Facultative anaerobic lactobacilli which produce lactic acid, alcohol (ethanol) or acetic acid and carbon dioxide from carbohydrates. As with other lactic acid bacteria these transient lactobacilli encourage, because of their formation of lactic acid, etc., a more acid environment in which there is a strong inhibition of less desirable micro-organisms.
2. *Lactobacillus acidophilus*	A natural inhabitant of the human small and large intestine. Found in the intestine of humans and animals, human mouth and vagina. Facultative anaerobic lactobacilli (grow in the presence or absence of air) which produce lactic acid as a main product from carbohydrates. Optimum growth temperature 35-38°C (95-100°F). The major beneficial functions of acidobacteria are: (a) They enhance and allow digestion of milk sugar (lactose) by producing the enzyme lactase; and generally aid in the digestion of nutrients. (b) They are able by some competitive

	means, e.g. creation of lactic acid and other inhibitory substances, to suppress undesirable micro-organisms in the intestine. (c) Some strains act to help destroy hostile invading bacteria by producing natural antibiotic substances. (d) Some strains are able to help reduce the level of cholesterol thus lessening the dangers of cardiovascular complications. (e) They are able to help lessen the proliferation of hostile yeasts such as *Candida albicans*. When the intestinal microflora is disturbed (the lactobacilli can be adversely affected) under the influence of oral antibiotic therapy, or stress conditions, the use of supplemental *acidophilus*, in food or concentrated form, can reverse such negative processes. The regular use of *acidophilus* bacteria as a supplement or in food is a protective means against an imbalance of the intestinal microflora.
3. *Lactobacillus bulgaricus*	A transient yet important bacteria in human ecology. Together with *Streptococcus thermophilus* constitutes a yogurt culture which is used in the preparation of yogurt. Found in yogurt and cheese. Facultative anaerobic lactobacilli which produce lactic acid as a main product from carbohydrates. Optimum growth temperature 40–43 °C (104–109 °F). Some strains produce natural antibiotic substances. When eaten or taken supplementally these bacteria benefit humans by enhancing digestion of milk sugar by producing the enzyme lactase. As with other lactic acid bacteria, these transient bacteria encourage, because of

	their creation of lactic acid, a more acid environment in which there is a strong inhibition of less desirable micro-organisms.
4. *Lactobacillus plantarum*	A transient bacteria of the human intestine. Found in dairy products, sauerkraut, pickled vegetables, silage, cow dung and the human intestine and stools. Facultative anaerobic lactobacilli which produce lactic acid as a main product from carbohydrates. Optimum growth temperature 30–35°C (86–95°F). As with other lactic acid bacteria they encourage, because of their formation of lactic acid, a more acid environment in which there is a strong inhibition of less desirable micro-organisms.
5. *Lactobacillus casei* subspecies *rhamnosus*	A transient bacteria of the human intestine. Found in milk and cheese, dairy products, cow dung, silage, and the human intestine and mouth. Facultative anaerobic lactobacilli which produce lactic acid as a main product from carbohydrates. Optimum growth temperature around 37°C (98.6°F). As reported, these bacteria may be effective in the treatment of certain intestinal conditions; however, more evidence is needed.
6. *Lactobacillus brevis*	A transient bacteria of the human intestine. Found in milk, kefir, cheese, sauerkraut, silage, cow manure and the intestine and mouth of humans. Facultative anaerobic lactobacilli which produce lactic acid, alcohol (ethanol) or acetic acid and carbon dioxide from carbohydrates. Optimum growth at about 30°C (86°F). Some strains produce natural antibiotic substances.

	Evidence is lacking about the beneficial functions of these bacteria in the intestine; except that as with other lactic acid bacteria they encourage, because of their formation of lactic acid, a more acid environment which is inhibitory to less desirable micro-organisms.
7. *Lactobacillus salivarius*	A natural resident of the human intestine. Found in the mouth and intestinal tract of humans and hamsters, and intestinal tract of chickens. Facultative anaerobic lactobacilli which produce lactic acid as a main product from carbohydrates. Optimum growth at 35–40 °C (95–104 °F). As with other lactic acid bacteria they encourage, because of their creation of lactic acid, a more acid environment in which less desirable micro-organisms are inhibited. More evidence is needed about the beneficial functions of these bacteria in the human intestine.
8. *Lactobacillus delbrueckii* subspecies *delbrueckii*	A transient bacteria of the human intestine. Found in plant material, e.g. grain and vegetable mashes fermented at high temperatures (40–53 °C/104–127 °F). Facultative anaerobic lactobacilli which produce lactic acid from carbohydrates (unable to ferment milk sugar). Optimum growth at 40–44 °C (104–110 °F). Evidence is lacking about the possible beneficial functions of these bacteria in the human intestine.
9. *Lactobacillus caucasicus*	New designation: *Lactobacillus kefir* (see under *L. kefir*).
10. *Lactobacillus yoghurti*	Valid name: *Lactobacillus jugurti*, strains which do

(Lactobacillus jugurti)	not ferment sugar maltose. A transient bacteria of the human intestine. Found in sour milk, cheese (particularly Emmenthal and Gruyère), and cheese starter. Facultative anaerobic lactobacilli which produce lactic acid as a main product from carbohydrates. High acid producers, up to 2 per cent lactic acid. Optimum growth temperature 20–42°C (104–108°F). Evidence is lacking about the possible beneficial functions in the human intestine; except that as with other lactic acid bacteria these bacteria encourage, because of their formation of lactic acid, a more acid environment which is inhibitory to less desirable micro-organisms.
11. *Bifido-bacterium bifidum*	A natural inhabitant of the human intestine, but also found in the human vagina. Occur in larger numbers in the large intestine than in the lower part of the small intestine. *B. bifidum* together with other bifidobacteria species are the predominant organisms in the large intestine of breast-fed infants accounting for about 99 per cent of the microflora. In adolescents and adults, bifidobacteria are a major component of the large intestine's microflora. The levels of bifidobacteria decline with age and also in various conditions of ill-health. They produce acetic and lactic acids, with small amounts of formic acid from fermentable carbohydrates. Anaerobic bacteria with optimum temperature for growth 37–41°C (98.6–106°F). The major beneficial functions of bifidobacteria are: (a) The prevention of the colonization of the intestine by invading pathogenic bacteria or yeasts with which they

	compete for nutrients and attachment sites. (b) The production of acetic and lactic acids which lower the pH (increase the acidity) of the intestine thus making the region undesirable for other possibly harmful bacteria. (c) Assisting in nitrogen retention and weight gain in infants. (d) The inhibition of bacteria which can alter nitrates in the intestine (derived from food or water) into potentially harmful nitrites. (e) The production of B vitamins. (f) Assisting in the dietary management of liver conditions. When the intestinal microflora is disturbed (and consequently bifidobacteria decline) under the influence of oral antibiotic therapy, therapeutic irradiation of the abdomen, reduced gastric acidity, impaired intestinal motility, stresses or some other condition, bifidobacteria supplements or bifidobacteria found in food products such as bifidus milk can help to restore the intestinal microflora.
12. *Bifidobacterium infantis*	A natural inhabitant of the intestine of human infants, but also occurs in small numbers in the human vagina. Together with other bifidobacteria such as *B. bifidum, B. longum, B. breve* they are the predominant bacteria in the large intestine of infants. Anaerobic bacteria which produce acetic and lactic acids and small amounts of formic acid from carbohydrates. The major beneficial functions are similar to those of *B. bifidum*, as: (a) The prevention of colonization of the

	intestine by invading pathogens through competition for nutrients and attachment sites. (b) The production of acetic and lactic acids which increase the acidity of the intestine and thereby inhibit undesirable bacteria. (c) Assisting in nitrogen retention and weight gain in infants. (d) The inhibition of bacteria which can alter nitrates to potentially harmful nitrites. (e) The production of B vitamins. When the microflora of infants is disturbed (and the levels of bifidobacteria decline) under the influence of sudden changes in nutrition, use of antibiotics, vaccination, convalescence or sudden weather changes, the use of supplemental bifidobacteria, or in food, can help in the nutritional restoration of the intestinal microflora.
13. *Bifido-bacterium longum*	A natural inhabitant of the human intestine. Found in the stools of human infants and adults. Together with bifidobacteria such as *B. bifidum, B. infantis* and *B. breve* they are the predominant bacteria in the large intestine of infants. A separate biotype of *B. longum* occurs in large numbers in the large intestine of adolescents and adults. Anaerobic bacteria which produce acetic and lactic acids with small amounts of formic acid from carbohydrates. A wider range of carbohydrates fermented compared with that of *B. bifidum*. The beneficial roles are similar to those of *B. bifidum*, as: (a) The prevention of colonization of the intestine by invading pathogens with

	which they compete for nutrients and attachment sites. (b) The production of organic acids which increase the acidity of the intestine and thereby inhibit undesirable bacteria. (c) Assisting in nitrogen retention and weight gain in infants. (d) The inhibition of bacteria which alter nitrates to potentially harmful nitrites. (e) The production of B vitamins. If the intestinal microflora is disturbed (and the levels of bifidobacteria decline) under the influence of antibiotics, irradiation of the abdomen with gamma or x-rays, reduced gastric acidity, or stress conditions, the use of supplemental bifidobacteria, or bifidobacteria found in food products, can help in the restoration of the microflora.
14. *Bacillus subtilis*	Aerobic sporeforming bacteria found in the human intestine in very small numbers. Usually found in a variety of sources. Causative agent of ropy (slimy) bread. They have been implicated in food poisoning, and therefore may be regarded as undesirable bacteria.
15. *Bacillus sphaericus*	Aerobic sporeformers which occur in soil, marine and freshwater sediment, milk and foods. Reported to be implicated in food poisoning, and therefore may be regarded as undesirable bacteria.
16. *Bacillus laterosporus*	Aerobic sporeforming bacteria, sometimes found in the human intestine in very small numbers. No evidence of possible beneficial effects in the human intestine. Any such effects may be questioned because these bacteria are not lactic acid producers, and occur in very small proportions.

17. *Streptococcus thermophilus*	A transient, i.e. non-resident bacteria of the human intestine. Together with *Lactobacillus bulgaricus* constitutes a yogurt culture which is used in the preparation of yogurt. Found in yogurt and cheese, heated and pasteurized milk. Facultative anaerobic streptococci which produce lactic acid as a main product from fermentable carbohydrates. Some strains produce antibiotic-like substances. Optimum growth temperature 40–45 °C (104–113 °F). As with other lactic acid bacteria these transient bacteria encourage, because of their creation of lactic acid, a more acid environment which is inhibitory to less desirable bacteria. When eaten or taken supplementally these bacteria benefit humans by enhancing digestion of milk sugar by producing the enzyme lactase. These bacteria are the only streptococci which produce lactase; and they produce it even more than *Lactobacillus bulgaricus*.
18. Streptococcus faecium	A resident bacteria of the human intestine, occuring in small numbers. Sources: faeces of humans and animals; insects and plants. Usually found in non-sterile foods. Facultative anaerobic streptococci, called enterococci which produce lactic acid as a main product from carbohydrates. Used as a component of starters (in some dairies) for the manufacture of some kinds of cheese. They are resistant to unfavourable environmental conditions, including acidity, temperature, salt, drying, atmospheric oxygen, etc. Scarce scientific evidence about the possible beneficial functions of these bacteria in the human

	intestine; especially when compared with *L. acidophilus* and bifidobacteria, and even with transient yogurt bacteria (*L. bulgaricus* and *S. thermophilus*).
19. *Streptococcus faecalis*	A resident bacteria of the human intestine, occurring in small numbers. Sources: faeces of humans and animals; insects and plants. Facultative streptococci, called enterococci which produce lactic acid from carbohydrates; some biotypes digest casein and gelatin. They are resistant to unfavourable environmental conditions. They can produce higher than normal levels of amines such as tyramine and histamine in the cheese; and tyramine, for example, may be involved in the onset of migraine attacks in susceptible persons. Normally, these bacteria are not used as a starter for the manufacture of fermented milk products. There is no firm evidence about the possible beneficial effects of these bacteria for humans. As reported, they may be pathogenic agents in urinary tract infections.
20. *Streptococcus lactis*	A transient bacteria of the human intestine. Sources: raw milk and milk products, plants and non-sterile frozen and dry foods. Facultative anaerobic streptococci which produce lactice acid as a main product from carbohydrates. Used as a starter for the manufacture of cheese, particularly cheddar, and as a component of starter for the manufacture of cottage cheese and cultured buttermilk. Some strains produce antibiotic nisin. The beneficial functions of these transient bacteria in the human intestine are comparatively less than those of

	acidophilus and bifidobacteria, as well as yogurt bacteria.

References

Abdel-bar, N.M. and Harris, N.D., (1984) 'Inhibitory effect of *Lactobacillus bulgaricus* on psychotrophic bacteria in associative cultures and in refrigerated foods', *J. Food Protection* 47, 61-64.

Agel, E.N., Friend, B.A., Long, C.A. and Shahani, K.M., (1982) 'Bacterial content of raw and processed human milk', *Journal of Food Protection*, 45, No. 6, 533-536, (April).

Alm, L. et al., (1983) 'The effect of acidophilus milk in the treatment of constipation in hospitalized geriatric patients', *XV Symp. Seed. Nutr. Found*.

Alm, Livia, (1982) 'Effects of fermentation on curd size and digestibility of milk proteins in vitro of Swedish fermented products', *Journal Dairy Science*, 64 (4) 509-514.

Alm, L., (1982) 'Effect of fermentation on B-vitamin content of milk in Sweden', *J. Dairy Sci*. 65, 353-359.

Alm, L., (1983) 'The effect of *Lactobacillus acidophilus* administration upon the survival of salmonella in randomly selected human carriers', *Prog. Fd. Nutr. Sci.* 7, 13-17.

Anand, S.K., Srinivasan, R.A. and Rao, L.K., (1984) Antibacterial activity associated with Bifidobacterium bifidum', *Cultured Dairy Products J.* 19, 6-8.

Archer, D.L. et al., (1985) 'Intestinal infection and malnutrition initiate Acquired Immune Deficiency Syndrome (AIDS)', *Nutrition Research* 5, 9-19.

Ayebo, A.D. et al., 'Antitumor component(s) of yogurt, I. Fractionation', Published as Paper No. 6142, Journal Series, *Nebraska Agricultural Experiment Station*.

Batish, V.K., Chander H. and Ranjanathan, B., (1985) 'Factors affecting enterotoxin production by thermonuclease positive streptococcus faecium IF-100 isolated from an infant food', *Journal of Food Science*, 50, p. 1513–1514.

Beck, C. and H. Necheles, (1961) 'Beneficial effects of administration of *Lactobacillus acidophilus* in diarrheal and other intestinal disorders', *American J. Gastroenterology* 35, 522–533.

Beerens, H., Romond, C. and Neut, C., (1980) 'Influence of breast feeding on bifido flora of newborn intestine', *Amer. J. Clin. Nutr*. 33, 2434–2439.

Bellomo, G. et al., (1979). 'New Prospects in the Treatment of Enteritides in Paediatrics'. *Medicine et Hygiene* 37, 3781–4. (31 Oct.).

Bellomo, G. et al., (1980). 'Controlled Double-blind Study of SF68 for Treatment of Diarrhoea in Paediatrics'. *Current Therapeutics Research* 28, 827–936.

Bogdanov, I.G. and Dalev, P.G. (1975) 'Antitumour glycopeptides from *Lactobacillus bulgaricus* cell wall', *FEBS Letters* 57, 259–261.

Bryant, M., (1986) 'The shift to probiotics', *J. Alternative Medicine* (Feb), 6–9.

Bullen, C.L., Tearle, P.V. and Willis, A.T., (1976) 'Bifidobacteria in the intestinal tract of infants: an in vivo study', *J. Med. Microbiol*. 9, 335–344.

Butler, B.C. and Beakley, J.W., (1960) 'Bacterial flora in vaginitis', *Am. J. Obst. & Gynec*. 79, 432–440.

Camarri, E. et al., (1981). 'A Double Blind Comparison of Two Different Treatments for Acute Enteritis in Adults'. *Chemotherapy* 27, 466–470.

Carlson, E., (1983) 'Enhancement by Candida of *S. aureus, S. marcescens, S. faecalis* in the establishment of infection', *Infection and Immunity*, 39:1, 193–197, (Jan.).

Collins, D., (1986) 'Colon Therapy', *Textbook of Natural Medicine*, (Ed. Pizzorno and Mrray), JBCNM, Seattle.

Collins, E.B. and Pamela Hardt, (1980) 'Inhibition of *Candida albicans* by *Lactobacillus acidophilus*', *Journal of Dairy Science*, 5, 830–832, (May).

Crawford, J.T., *Microbiology*, Chapter 56: 'Microbiota of dental plaque, caries, gingival disease, related systemic infections', 16th edition.

Del Vecchio-Blanco, C. et al., (1981). 'Effect of Treatment with SF68 on the Blood Ammonia Curve Following Protein Loading in Subjects with Heptic Cirrhosis'. *Medicine et Hygiene* 39, 2237-2389.

Donovan, P., (1986) 'Bowel Toxaemia, Permeability and Disease', *Textbook of Natural Medicine* (Ed. Pizzorno and Murray), JBCNM, Seattle.

Dowson, D., (1988) 'Colonic Cleansing', *Jn. Alternative and Complementary Medicine*, 31-33, (May).

Dubos, R. and Schaedler, R.W., (1962) 'Some biological effects of the digestive flora', *Amer. J. Med. Sciences* (Sep), 265-271.

Fernandes, C.F., Shahani, K.M. and Amer, M.A., (1987) 'Therapeutic role of dietary lactobacilli and lactobacillic fermented dairy products', *FEMS Microbiology Reviews* 46, 343-356.

Fernandes, C.F., Shahani, K.M. and Amer, M.A., (1988) 'Control of diarrhea by lactobacilli', *J. Appld. Nutr.* 40, 32-42.

Ferreira, C.L. and Gilliland, S.E., (1988) 'Bacteriocin involved in premature death of *Lactobacillus acidophilus* NCFM during growth at pH 6', *J. Dairy Sci*. 71, 306-315.

Finegold, S.M. et al., 'Comparative effect of broad-spectrum antibiotics on non-sporeforming anaerobes and normal bowel flora', *Annals New York Academy of Sciences* 145, 269-281.

Fischer, W.L., (1987) *How to Fight Cancer and Win*. Alive Books, Vancouver, Canada.

Fishbaugh, E., (1929) 'The colon in relation to chronic arthritis', *Am. J. of Surgery*, 7: 561-567.

Friend, B.A. and Shahani, K.M., (1984) 'Nutritional and therapeutic aspects of Lactobacilli, *J. Appld. Nutr*. 36, 125-153.

Giannella, R.A. et al., (1972) 'Gastric acid barrier to ingested micro-organisms in man: studies in vivo and in vitro', *Gut* 13, 251-256.

Gilbert, J.P. et al., (1983) 'Viricidal effects of lactobacillus and yeast fermentation', *Appl. Environ. Microbiol.* 46, 452-458.

Gilliland S.E. and Speck, M.L., (1977) 'Antagonistic action of *Lactobacillus acidophilus* toward intestinal and foodborne pathogens in associative cultures', *J. Food Protection* 40, 820-823.

Gilliland, S.E. et al., (1985) 'Assimilation of cholesterol by *Lactobacillus acidophilus*', *Applied and Environmental Microbiology* (Feb), 377-381.

Gilliland, S.E., (1979) 'Beneficial interrelationships between certain micro-organisms and humans: candidate micro-organisms for use as dietary adjuncts', *J. Food Protection* 42, 164–167.

Gilliland, S.E. and Speck, M.L., (1977) 'Instability of *Lactobacillus acidophilus* in yogurt', *J. Dairy Sci.* 60, 1394–1398.

Goldin, B.R., Seson, L., Dwyer, J., Sevfon, M. and Gorbach, L., (1980) 'Effect of diet and *Lactobacillus acidophilus* supplements on human fecal bacterial enzymes', *Nat'l. Cancer Institute*, 64, 255–261.

Gorbach, S.L., Chang, T. and Goldin, B., (1987) 'Successful treatment of relapsing clostridium difficile colitis with Lactobacillus GG', *Lancet*, 26 Dec., 1519.

Gracey, M.S., (1981) 'Nutrition, bacteria and the gut', *Br. Med. Bull.* 37, 71–75.

Grutte, F. and Muller-Beuthow, W., (1980) '*Human Gastrointestinal Microflora*, 'Instability of the normal intestinal flora in human infants', 39–44, J.A. Barth, Verlag, Leipzig.

Grutte, F.K. and Muller-Beuthow, W., (1980) 'Instability of the normal intestinal flora in human infants', *Human Gastrointestinal Microflora*, pp: 39–44, J.A. Barth, Verlag Leipzig.

National Dairy Council (1979) 'Gut ecology and health implications', *Dairy Council Digest* 50, 13–17.

Hamdan, I.Y. and Mikolajcik, E.M., (1974) 'Acidolin: an antibiotic produced by *Lactobacillus acidophilus*', *J. Antibiotics* 8, 631–636.

Hangee-Bauer, C.S., 'Lactobacilli and human health', Rounds/Journal Club of the John Bastyr College of Naturopathic Medicine, Seattle, Washington; 335–340.

Hansen, Robert, 'Bifidobacteria have come to stay'. (1985) Reprint from *North European Dairy Journal*, No. 3, 2–6.

Hargrove, R.E. and Alford, J.A., (1978) 'Growth rate and feed efficiency of rats fed yogurt and other fermented milks', *Journal of Dairy Science*, 61, (1) 11–19, 1978.

Henteges, D.J. et al., (1977) 'Effect of high-beef diet on the fecal bacterial flora of humans', *Cancer Research* 37, 568–571.

Hepner, G. et al., (1979) 'Hypocholesterolemic effect of yogurt and milk', *Amer. J. Clin. Nutr.* 32, 19–24.

Huppert, M. et al., (1953) 'Pathogenesis of *Candida albicans*; I. The effect of antibiotics, on the growth of *Candida albicans*', 65, 171–176.

Jackson, G.G. and Dowling, H.F., (1953) 'Adverse effects of antibiotic treatment', *G.P.* (Aug), 34–40.

Jameson, R.M., (1976) 'The prevention of recurrent urinary tract infection in women', *The Practitioner* 216, 178-181.

Kageyama, T. et al., (1984) 'The effect of bifidobacterium administration in patients with leukemia', *Bifidobacteria Microflora*, 3, 29-33.

Kantor, J. et al., (1930) 'Colon Studies, cecal stasis; its significance to proximal colon stasis', *Amer. Jn. Roentgenology and Radium Therapy*, 24(1): 1-20.

Kay, H.W. and Heeschen, W., (1983) 'Antileukemic effects in mice from fermentation products of *Lactobacillus bulgaricus*', *Milchwissenschaft* 38, 257-260.

Kilara, A. and K.M. Shahani, (1976) 'Effect of cryoprotective agents on freeze-drying and storage of lactic cultures', *Cultured Dairy Products Journal*, 11 (5) 8-11.

Kilara, A. and K.M. Shahani, (1987) 'Lactic fermentations of dairy foods and their biological significance', *Journal Dairy Science*, 61, 1793-1800.

Kim, J.S. and Gilliland, S.E., (1983) '*Lactobacillus acidophilus* as a dietary adjunct for milk to aid lactose digestion in humans', *J. Dairy Sci*. 66, 959-966.

Kim, H.S., (1988) 'Characterization of lactobacilli and bifidobacteria fats applied to dietary adjuncts', *Cultured Dairy Products J.* (Aug), 6-9.

Law, B.A. and Sharpe, M.E. (1978) *Streptococci*, 'Streptococci in the dairy industry', 263-278, Academic Press Inc., London.

LeRoith, D., Shiloach, J., Roth, J. and Lesniak, M., (1981) 'Insulin or a closely related molecule is native to *E. coli*', *Jn Biol-Chemistry*, 256: 6533-6536.

Mann, G.V. and A. Sperry, (1974) 'Studies of a surfactant and cholesteremia in the Masai', *American Journal Clinical Nutrition*, 27, 464-469.

Marshall, H. and Thomson, C., (1932) 'Colon irrigation in the treatment of mental disease', *N. Eng. J. of Med.*, 207: 454-457.

Matalon, M.E. and Sandine, W.E., (1986) '*Lactobacillus bulgaricus, Streptococcus thermophilus* and yogurt: a review, *Cultured Dairy Products J.* Nov, 6-11.

Mikolajcik, E.M. and Hamdan, I.Y., (1975) '*Lactobacillus acidophilus*, II. Antimicrobial agents', *Cultured Dairy Products J.* (Feb). 18-20.

Moroni, M., (1979). *Vecchi E. Nuovi Orientamenti nel rattamento delle enteriti Medico epaziente* (March).

Mott, G.E. et al., (1973) 'Lowering of serum cholesterol by intestinal bacteria in cholesterol-fed piglets', *Lipids* 8, 4282–431.

Muting, D. et al., (1968) 'The effect of bacterium bifidum on intestinal bacterial flora and toxic protein metabolites in chronic liver disease', *Am. J. Proctology* 19, 336–342.

Ninkaya, R., (1984) 'Role of bifidobacteria in enteric infection', *Bifidobacteria Microflora* 5, 51–55.

Nikolov, N.M. (1962) 'Acidophilus paste, Bulgarian yogurt and other cultured milk products', *Zemisdat* — Sofia, Bulgaria.

Okamura, N. et al., (1986) 'Interaction of shigella with bifidobacteria', *Bifidobacteria Microflora* 5, 51–55.

Pabst, M.J. et al., (1982) 'Cultured human monocytes require exposure to bacterial products to maintain an optimal oxygen radical response', *J. Immunology* 128, 123–128.

Pearce, J.L. and Hamilton, J.R., (1974) 'Controlled trial or orally administered lactobacilli in acute infantile diarrhea', *J. Pediatrics* 84, 261-262.

Perdigon, M.E. et al., (1987) 'Enhancement of immune response in mice fed with *Streptococcus thermophilus* and *Lactobacillus acidophilus*.

Pollman, D.S., Danielson, D.M., Wren, W.B., Peo, E.R. Jr. and Shahani, K.M., (1980). 'Influence of *Lactobacillus acidophilus* on gnotobiotic and conventional pigs', *Journal of Animal Science*, 51, (3), 629–637.

Poupard, J.A., Husain, I. and Norris, R.F., (1973) 'Biology of the Bifidobacteria', *Bacteriological Reviews*, 37, (2), 136–165 (June).

Quin, T.C., (1984) 'Gay bowel syndrome, the broadened spectrum of nongenital infection', *Postgraduate Medicine* 76, 197–210.

Rao, D.R. and Shahani, K.M., (1987) 'Vitamin content of cultured milk products', *Cultured Dairy Products J.* Feb, 6–10.

Rapoport, L. and Levine, W.I., (1965) 'Treatment of oral ulceration with lactobacillus tablets, *Oral Surgery, Oral Medicine and Oral Patholol*g*y* 20, 591–593.

Rašić, J.L., (1988) 'Occurrence of *B. infantis* and *B. bifidum* in the gut of infants and adults', Letter of correspondence.

Rašić, J.L. and Kurmann, J.A., (1983) *Bifidobacteria and Their Role*, Birkhauser Verlag, Boston.

Rašić, J.L., (1983) 'The role of dairy foods containing bifido and acidophilus bacteria in nutrition and health', *North European Dairy J.* 4, 80–88.

Rašić, J.L., (1987) 'Nutritive value of yogurt', *Cultured Dairy Products J.* (Aug), 6-9.

Reddy, G.V. et al., (1983) 'Antitumour activity of yogurt components', *J. Food Protection* 46, 8-11.

Reddy, G.V., Shahani, K.M., Friend, B.A. and Chandan, R.C., (1983) 'Natural antibiotic activity of *L. acidophilus* and *bulgaricus*', *Cultured Dairy Products*, 18, (2), 15-19.

Rhoads, J.L. et al., (1987) 'Chronic vaginal candidiasis in women with human immunodeficiency virus infection', *JAMA* 257, 3105-3107.

Riise, T., (1981) 'The probiotic concept — a review', *Chris Hansen's Laboratory*, Copenhagen, Denmark, (June), 1-8.

Robbins-Browne, R.M. and Levine, M.M., (1981) 'The fate of ingested lactobacilli in the proximal small intestine, *Am. J. Clin. Nutr.* 34, 514-519.

Rowland, I.R. and Grasso, P., (1975) 'Degradation of N-Nitrosamines by intestinal bacteria', *Applied Microbiology* (Jan), 7-12.

Savage, D.C. (1987) 'Factors influencing biocontrol of bacterial pathogens in the intestines', *Food Technology* (July), 82-87.

Schecter, A., Ryan, J. Constable, J., (1987) 'Polychlorinated PCDD and PCDF in human breast milk from Vietnam compared with cow's milk and human milk from North American Continent', *Chemosphere* 16, 2003-2016.

Schecter, A., Gasiewicz, T., (1987) *Solving Hazardous Waste Problems*, Chapter 12, 'Human breast milk levels of dioxins and dibenzofurans: Significance with respect to current risks', Exner. American Chemical Society, Washington D.C.

Schecter, A. and Gasiewicz, T., (1987) 'Health hazard assessment of chlorinated dioxins and dibenzflurans contained in human milk', *Chemosphere* 16, 2147-2154.

Shahani, K.M. and Chandan, R.C., (1979) 'Nutritional and healthful aspects of cultured and culture-containing diary products', *J. Dairy Sci.* 62, 1685-1694.

Shahani, K.M., (1987) 'Antiviral and antifungal effect of lactobacilli', *The New Show Daily*, Sunday 12 July, 1987.

Shahani, K.M. et al., (1980) 'Role and significance of enzymes in human milk', *Amer. J. Clin. Nutr.* 33, 1861-1868.

Shahani, K.M. and Ayebo, A.D., (1980) 'Role of dietary lactobacilli in gastrointestinal microecology', *American Journal of Clinical Nutrition*, 33, 2248-2257 (Nov.)

Shahani, K.M. and Friend, B.A., (1973) 'Nutritional and therapeutic aspects of Lactobacilli', *Journal Appl. Nutr.* 37, (2), 136-165, (June).

Shahani, K.M. and Vakil Jayantkumar, R., (1972) 'Antibiotic acidophilus and process of preparing the same', United States Patent No. 3,689,640, 5 Sept. 1972.

Shahani, K.M., Reddy, G.V. and Joe, A.M., (1974) 'Nutritional and Therapeutic aspects of cultured dairy products', *Proc XIX, Inter'l. Dairy Cong.,* Ie, 569-570.

Shahani, K.M. Vakil, J.R., and Kilara, A., (1977) 'Natural antibiotic activity of Lactobacillus acidophilus and bulgaricus II Isolation of acidophilin from *L. acidophilus*', *Cultured Dairy Products Journal*, 12, (2) 8-11.

Simon, G.L. and Gorbach, S.L., (1981) 'Intestinal flora in health and disease', *Physiology of the Gastrointestinal Tract*, edited by L.R. Johnson, pp. 1361-1380, Raven Press, New York.

Siver, R.H., (1964) 'Lactobacillus for the control of acne', *J. Medical Society New Jersey* 58, 52-83.

Sneath, P.H.A., (1986) *Bergey's Manual of Systematic Bacteriology,* Volume 2. Williams & Wilkins, Baltimore.

Socransky, S.S., (1977) 'Bacteriological studies of developing supragingival dental plaque', *J. Periodontal Res.* 12, 90-106.

Speck, M.L., (1975) 'Interactions among lactobacilli and Man', *J. Dairy Sci*. 59, 338-343.

Speck, M.L., (1975) 'Contributions of micro-organisms to food and nutrition', *Nutrition News*, 38, (4), 13, (Dec.)

Thompson, L.U., David, J.A., Jenkins, M.A. and Amer, V. et al., (1982) 'The effect of fermented and unfermented milks on serum cholesterol', *American Journal Clin. Nutr.,* 36, 11, 6, 111, (Dec).

United States Government, Porubcan, et al., United States Patent 4,115,119, 'Preparation of culture concentrates for direct vat set cheese production', 19 Sept. 1978.

Vos, J.G., (1977) 'Immune suppression as related to toxicology', *CRC Critical Reviews in Toxicology*, 67-101.

Warshaw, A., Bellini, C. and Walker W., (1977) 'The intestinal mucosal barrier to intact antigen protein', *American J. Surgery*, 133: 55-58.

Warshaw, A., Walker, W., Cornell, W. and Isselbacher, K., (1971) 'Small intestine permeability to macromolecules', *Lab. Invest.*, 25: 675-684.

Warshaw, A., Walker, W. and Isselbacher, K., (1974) 'Protein uptake in the intestines; Evidence of intact macromolecules', *Gastroenterology*, 66: 987–992.

Weekes, D.J., (1983) 'Management of Herpes Simplex with a virostatic bacterial agent', *E.E.N.T. Digest* 25.

Wheater, D.M. et al., (1952) 'Possible identity of "Lactobacillin" with hydrogen peroxide produced by Lactobacilli', *Nature* 170, 623–624.

Williams, E. and Hemmings, W., (1979) 'Intestinal uptake and transport of proteins in the adult rat', *Proc. Royal Soc. London*, 203: 177–189.

Zaika, L.L. et al., (1983) 'Inhibition of lactic acid bacteria by herbs', *J. Food Science* 48, 1455–1459.

Bibliography

Bland, Jeffrey, *Yeast nutrition — Candida albicans: An unsuspected problem*, Nutritional Biochemistry, University of Puget Sound, Tacoma, Washington.

Bogdanov, I., *Observations on the therapeutic effect of the anti-cancer preparation from Lactobacillus bulgaricus (LB-51) tested on 100 oncological patients*. Laboratory for the Research and Production of Biologically Active Substances, Sofia, Bulgaria (1982).

Chaitow, Leon, *Candida Albicans: Could Yeast be Your Problem?* Thorsons, Wellingborough (1985).

Chaitow, Leon and Martin, Simon, *A World Without AIDS*, Thorsons, Wellingborough (1988).

Crooke, William G., *The Yeast Connection,* Vintage Books (1988).

Culbert, Michael, *AIDS: Terror, Truth, Triumph,* Bradford Foundation (1987).

Dubos, René, *Man Adapting*, Yale Press (1976).

Helferich, W.M. and Westhoff, D., *All About Yogurt*, Prentice Hall Inc. (1980).

Metchnickoff, E., *The Prolongation of Life*, G.P. Putnam & Sons, New York (1908).

Mindell, Earl, *Unsafe at any Meal*, Warner Books, New York, 1987.

Pizzorno, J. and Murray, M., *Textbook of Natural Medicine*, JBCNM, Seattle (1987).

Rašić, Jeremija and Kurmann, Joseph, *Bifidobacteria and Their Role*, Birkhauser Verlag, Basel (1983).

Seaman, B. and G., *Women and the Crisis in Sex Hormones*, Harvester Press (1978).

Stein, Jay (ed.), *Internal Medicine*, Little-Brown, Boston (1983).

Truss, C. Orion, *Missing Diagnosis*, C. Orion Truss, Birmingham, Alabama (1982).

Werbach, Melvyn, *Nutritional Influences on Illness*, Thorsons, Wellingborough (1989).

Index